VISUAL CULTURE AND ARCTIC VOYAGES

In the mid-nineteenth century, thirty-six expeditions set out for the Northwest Passage in search of Sir John Franklin's missing expedition. The array of visual and textual material produced on these voyages was to have a profound impact on the idea of the Arctic in the Victorian imaginary. Eavan O'Dochartaigh closely examines neglected archival sources to show how pictures created in the Arctic fed into a metropolitan view transmitted through engravings, lithographs, and panoramas. Although the metropolitan Arctic revolved around a fulcrum of heroism, terror, and the sublime, the visual culture of the ship reveals a more complicated narrative that included cross-dressing, theatricals, dressmaking, and dances with local communities. O'Dochartaigh's investigation into the nature of the on-board visual culture of the nineteenth-century Arctic presents a compelling challenge to the 'man-versus-nature' trope that still reverberates in polar imaginaries today. This title is also available as Open Access on Cambridge Core.

EAVAN O'DOCHARTAIGH is a postdoctoral researcher funded by an SFI-IRC Pathway Programme Award at National University of Ireland Galway. Prior to this she was a Marie Skłodowska-Curie Individual Fellow at Umeå University in northern Sweden and a Government of Ireland Doctoral Scholar at National University of Ireland Galway. She has also worked as an archaeologist and archaeological illustrator in Ireland, Iceland, and the UK.

CAMBRIDGE STUDIES IN NINETEENTH-CENTURY LITERATURE AND CULTURE

Nineteenth-century literature and culture have proved a rich field for interdisciplinary studies. Since 1994, books in this series have tracked the intersections and tensions between Victorian literature and the visual arts, politics, gender and sexuality, race, social organisation, economic life, technical innovations, scientific thought – in short, culture in its broadest sense. Many of our books are now classics in a field which since the series' inception has seen powerful engagements with Marxism, feminism, visual studies, post-colonialism, critical race studies, new historicism, new formalism, transnationalism, queer studies, human rights and liberalism, disability studies and global studies. Theoretical challenges and historiographical shifts continue to unsettle scholarship on the nineteenth century in productive ways. New work on the body and the senses, the environment and climate, race and the decolonisation of literary studies, biopolitics and materiality, the animal and the human, the local and the global, politics and form, queerness and gender identities, and intersectional theory is re-animating the field. This series aims to accommodate and promote the most interesting work being undertaken on the frontiers of nineteenth-century literary studies, connecting the field with the urgent critical questions that are being asked today. We seek to publish work from a diverse range of authors, and stand for anti-racism, anti-colonialism and against discrimination in all forms.

A complete list of titles published will be found at the end of the book.

VISUAL CULTURE AND ARCTIC VOYAGES

Personal and Public Art and Literature of the Franklin Search Expeditions

EAVAN O'DOCHARTAIGH
National University of Ireland Galway

CAMBRIDGE
UNIVERSITY PRESS

CAMBRIDGE
UNIVERSITY PRESS

University Printing House, Cambridge CB2 8BS, United Kingdom

One Liberty Plaza, 20th Floor, New York, NY 10006, USA

477 Williamstown Road, Port Melbourne, VIC 3207, Australia

314–321, 3rd Floor, Plot 3, Splendor Forum, Jasola District Centre, New Delhi – 110025, India

103 Penang Road, #05–06/07, Visioncrest Commercial, Singapore 238467

Cambridge University Press is part of the University of Cambridge.

It furthers the University's mission by disseminating knowledge in the pursuit of education, learning, and research at the highest international levels of excellence.

www.cambridge.org
Information on this title: www.cambridge.org/9781108834339
DOI: 10.1017/9781108992794

First published 2022

A catalogue record for this publication is available from the British Library.

Library of Congress Cataloging-in-Publication Data
NAMES: O'Dochartaigh, Eavan, author.
TITLE: Visual culture and Arctic voyages : personal and public art and literature of the Franklin search expeditions / Eavan O'Dochartaigh, Umeå Universitet, Sweden.
DESCRIPTION: Cambridge ; New York, NY : Cambridge University Press, 2022. | Series: Cambridge studies in nineteenth-century literature and culture | Includes bibliographical references and index.
IDENTIFIERS: LCCN 2021051456 (print) | LCCN 2021051457 (ebook) | ISBN 9781108834339 (hardback) | ISBN 9781108994897 (paperback) | ISBN 9781108992794 (ebook)
SUBJECTS: LCSH: Franklin, John, 1786–1847 – Travel – Arctic regions. | Search and rescue operations – Arctic Ocean – History – 19th century. | Arctic regions – Discovery and exploration – British. | John Franklin Arctic Expedition (1845–1851) | Terror (Ship) | Erebus (Ship) | Northwest Passage – Discovery and exploration – British. | BISAC: LITERARY CRITICISM / European / English, Irish, Scottish, Welsh
CLASSIFICATION: LCC G625 .O36 2022 (print) | LCC G625 (ebook) | DDC 919.8–dc23/eng/20211214
LC record available at https://lccn.loc.gov/2021051456
LC ebook record available at https://lccn.loc.gov/2021051457

ISBN 978-1-108-83433-9 Hardback

Contents

Figures

Acknowledgements

I am grateful to the many people and institutions who have made this book possible. My academic interest in the Arctic was first encouraged by Michael Bravo at the Scott Polar Research Institute, University of Cambridge, and developed under the supervision of Daniel Carey (Department of English) and Nessa Cronin (Centre for Irish Studies) at National University of Ireland, Galway. Much of the work for the book was carried out as a PhD student, and I was extremely fortunate to benefit from their enthusiasm, intellectual rigour, and unfailing kindness. Further advice from Elizabeth Tilley and Geoff Quilley was crucial in revising the work for publication.

The research would not have happened without the Irish Research Council's Government of Ireland Postgraduate Scholarship, which funded the first three years, and the Hardiman Scholarship Fund at NUI Galway, which funded the final year of the doctorate. The Ireland Canada University Foundation awarded a Dobbin Scholarship funded by Steve Hudson, which enabled me to conduct crucial research at Library and Archives Canada, Ottawa. Bursaries from NUI Galway supported additional research, and a Gladstone Library Foundation scholarship gave me the opportunity to work at their beautiful residential library in Wales. More recently, a Marie Skłodowska-Curie Individual Fellowship (grant agreement no. 839477) at Umeå University, Sweden, funded by the European Commission under Horizon 2020, has facilitated revision and completion of the book. Additionally, this funding has covered the costs of photography, image permissions, gold open access, and professional indexing. Strategic resource funding from the Arctic Research Centre (Arcum) at Umeå University has also been put towards these costs.

Portions of the first and third chapters were published in an earlier form as 'Exceedingly Good Friends: The Representation of Indigenous People during the Franklin Search Expeditions to the Arctic, 1847–59' in *Victorian Studies* 61, no. 2 (2019); I appreciate the interest of Sarah Winter in connection with this. At Cambridge University Press, Bethany Thomas has been enthusiastic about this book right from the start and has been a pleasure to work with, as have George Paul Laver and the entire production and marketing team. I would also like to thank the series editors, Kate Flint and Clare Pettitt, the two anonymous reviewers whose comments greatly improved the book, and Madelon Nanninga-Franssen, who has created the index.

Throughout my research, I was aided by the staff of the James Hardiman Library, NUI Galway (particularly Rosie Dunne), the National Library of Ireland in Dublin, the National Maritime Museum in Greenwich, the Scott Polar Research Institute, Cambridge, and the Royal Geographical Society and British Library in London. Staff at the National Library of Australia, the Rubenstein Library in North Carolina, and the Norfolk Record Office helpfully answered queries and sent photographs. The exceptionally friendly and obliging staff at Library and Archives Canada in Ottawa made long days in the archives so much easier. Tim James, Julie Choudhury, and Alan Spencer provided homely accommodation during archival research in London and Ottawa.

I am appreciative of the support of many people who helped along the way, including staff at the Moore Institute for Research in the Humanities and Social Studies, the Centre for Irish Studies, and the English Department at National University of Ireland, Galway. I am so grateful to Rachel Hilliard for her thoughtful and insightful advice, for bringing Peta Freestone's 'Thesis Boot Camp' to Galway, and for her writing-group meetings, which are a continuing source of motivation. The spontaneous online chats with Méabh Ní Fhuartháin and Ursula Connolly as part of this group during the final stages of the book's completion were particularly welcome. Additionally, I would like to thank my mentor and dear friend Anne Karhio for her practical advice and many laughs along the way.

There are so many people to thank in Umeå, Sweden, where this book was finalised variously at the kitchen table, my daughter's homework desk, and an old sewing-machine table during the exceptionally snowy pandemic winter of 2020/1. I had the pleasure of being welcomed to Umeå University by Satish Strömberg, Ingela Westerlund, and Stefan Gelfgren, and I am grateful for the support of Maria Lindgren Leavenworth, Heidi Hansson, Peter Sköld, Anna-Stina Falkmer, and Daniel Andersson. I also need to thank Evelina Liliequist, Coppélie Cocq, Gavin Feller, Maria Stridsman, Bram Vaassen, and Gunilla Isaksson, as well as those who came to my social academic writing group in Umeå, both in person and then online, and helped to create a productive and friendly environment.

I have been extremely lucky to find myself part of the network for Ingeborg Høvik's *Arctic Voices* project based at UiT The Arctic University of Norway and funded by the Research Council of Norway. In particular, the encouragement, openness, and humour of Ingeborg, Sophie Gilmartin, and Sigfrid Kjeldaas have been a great source of academic motivation. I have met many inspiring, helpful, and friendly people at conferences and other academic events over the years, and I would especially like to thank Anna McLauchlan, Carl Thompson, Jean de Pomereu, Efram Sera-Shriar, Nanna Katrine Lüders Kaalund, Alessandro Antonello, Susanne Lalonde, Ingrid Urberg, Janicke Kaasa, Sarah Pickman, and Claire Warrior. In Ottawa, Chris Burn, Janice Cavell, and the Geography Department at Carleton University were very warm and welcoming, while in Kingston, Erika Behrisch Elce generously provided lunch and stimulating conversation. I made numerous friends as a PhD student at NUI Galway who each contributed in their own way, but Kathleen Pacious, Lucy Elvis, John Carrigy, Ciaran McDonough, Sarah Corrigan, Verena Commins, Isabel Corfe, and Mikyung Park deserve special mention.

I was fortunate to be amongst good friends at the Hólar archaeological project in Iceland when I received the news of the Irish Research Council award that would fund the doctoral research from which this book evolved. Many people listened as I worked through my decision to return to university, especially Ragnheiður Traustadóttir, Tara Carter, Jennica Einebrant Svensson, and Lísabet Guðmundsdóttir. Thanks are due to all my friends in Ireland and beyond for encouragement along the way (and care packages!), but especially Tanya Curry, Ruth Rowan, Aisling Barry, Mary O'Brien, Annie Atkins, and Board Game Club in Galway. In Umeå, I am grateful to all the lovely parents and kids (especially Eleonor and Jed Biolcati-Brennan, Maria Östensson, Javier Capella, and their families) who came to our outdoor meetings of Book Badgers and kept us sane during this pandemic winter with campfire chat and high-speed sledging. Our neighbours (and landlords) Karina DiLucia and Kent Sandström have been incredibly kind and generous to us and helped us in so many ways throughout our stay.

I would like to thank my siblings Aideen, Conor, and especially Niall, for valuable practical advice. During my PhD years, the care provided to my daughter by Susanna Groth, Caroline Forde, Rita Leonard, Ian MacLochlainn, and the O'Dochartaigh Barrs meant that I never had to worry. My parents Eoin and Niamh have been a constant and caring presence in my life. Finally, thank you to Svarten the pony for Christmas 2020, our two guinea pigs Mimi and Penguin for soothing munching noises, my husband Geoff for time, understanding, and unwavering support, and my hilarious daughter Sóley for running commentary, skiing lessons, and moving to northern Sweden.

Abbreviations and Nomenclature

BL	British Library, London
LAC	Library and Archives Canada, Ottawa
NLA	National Library of Australia, Canberra
NLI	National Library of Ireland, Dublin
NMM	National Maritime Museum, Greenwich, London
NRO	Norfolk Record Office, Norwich
RGS	Royal Geographical Society, London
RL	Rubenstein Library, North Carolina
SPRI	Scott Polar Research Institute, University of Cambridge
TNA	The National Archives, London
TPL	Toronto Public Library

The place names of the Arctic Regions that appear in historical records are generally different from those that are used today. In order to retain historical accuracy, when quoting from primary sources, I have retained peculiarities of spelling and used the historical place names given to the inlets, islands, and settlements of the Arctic by European explorers. Where possible, I have given the Indigenous or official name to settlements in parentheses. The word *Esquimaux* was commonly used in the nineteenth century to describe the Indigenous people of Greenland, Arctic Canada, and coastal Alaska. *Tuski* was used to refer to the Chukchi on the west side of the Bering Strait. Throughout this book, barring quotations, the Indigenous peoples of the Arctic are referred to as *Chukchi* (north-eastern Russia), *Yup'ik* and *Deg Hit'an* (western Alaska), *Iñupiat* (northern and western Alaska), *Gwich'in* (north-eastern Alaska and Gwich'in Settlement Area, Canada), *Inuvialuit* (Inuvialuit Settlement Region, Canada), *Inughuit* (north-western Greenland), and *Inuit* (Nunavut, Canada and Greenland). The term *Inuit* is also sometimes used to refer to *Iñupiat* and *Inuvialuit*, in accordance with the charter of the Inuit Circumpolar Council. As the term *Inuit* means 'the people', it is never prefaced with 'the'. In keeping with the linguistic recommendation from Government of Canada's Translation Bureau, *Inuk* is used as the singular noun, and *Inuit* is the plural noun and the adjective.

Introduction
Witnessing the Arctic

On Saturday, 9 February 1850, an eager crowd of invited guests entered the well-known Leicester Square rotunda in London to witness an immense new circular painting (or panorama[1]) created by Robert Burford and Henry Selous. As they mounted the steps to the raised viewing platform, a spectacular and vivid view of icebergs emerged before them. This was *Summer and Winter Views of the Polar Regions*, the latest, and arguably the finest, in a series of visual entertainments that claimed to show the little-known Arctic regions visited by a recently returned British naval expedition led by James Clark Ross. For weeks, adverts and letters in the papers had noted that this visual entertainment was distinguished by being the only one to be based on an actual eyewitness source: the on-the-spot sketches of William Henry James Browne, the Irish lieutenant who had served on the expedition. Thus, its accuracy and authenticity were well established long before its completion. With no external reference points on which to rest the eye, one woman, Lady Jane Franklin, was said to have spent two hours 'inspecting the picture', which must have had a 'peculiar interest' for her, given that it was 'near the place in which her husband and his expedition [were] supposed to be, if still alive'.[2]

It was almost five years since the large British naval expedition led by her husband, John Franklin (1786–1847), had disappeared into the Canadian Arctic.[3] Despite a huge effort by thirty-six search, supply, and relief expeditions between 1847 and 1859, no survivors of the Franklin expedition were ever found.[4] These expeditions represented a remarkable mobilisation of resources and men, funded by both the British government and private sponsors, including Jane Franklin herself and the American philanthropist and merchant Henry Grinnell; their combined endeavours kept the Arctic imaginary in the public eye throughout the 1850s.[5] During those voyages, shipboard expedition members produced large amounts of visual and written material for both scientific and sentimental reasons – topographical sketches, coastal profiles, maps, and written logs were produced alongside personal portraits, illustrated periodicals, travel sketches, and private journals. Although there were no professional artists on any of the search expeditions, in pictures, and in words, the Arctic experience was sketched and painted, described, and inscribed. In the metropole, too, following the return of the first search expedition in 1849, the Arctic was being re-imagined, engraved, lithographed, painted, and published. Above all, it was now being commodified and sensationalised. The Arctic was everywhere: at printsellers and panoramas, in published narratives and lithograph folios, in the press and in the theatre. *Visual Culture and Arctic Voyages* investigates how these representations of the Arctic came to be, with close attention to media and contexts, moving from the ship to the city, showing how the on-board Arctic was quite different to that of the metropole, and how this re-imagined Arctic continues to resonate in the twenty-first century.

By using close analysis of visual and verbal sources, I explore the processes of transformation of the representation of the Arctic from on-board sketches through to published texts, prints, and panoramas, enabling new conclusions to be drawn about the nature of the representation of the Arctic and of the Franklin searches. On the one hand, a narrative of humour, domesticity, familiarity, and of an often-benign environment permeates the on-board record. On the other, a battleground where man was pitted against nature dominated the public narrative.[6] In public archives alone, upwards of five hundred sketches, watercolours, and illustrations created by expedition members exist; many more must remain in private ownership. Over six hundred prints relating to the search appeared in published books, periodicals, and folios during the 1850s, while more than ten Arctic panoramas were exhibited in London. Other Arctic panoramas appeared in Ireland, Scotland, America, Canada, and Australia. Such pictures often created a specific visual narrative for public consumption that emphasised the laborious and difficult nature of the search amidst the 'savage horrors' of the polar regions.[7]

Although the Franklin expedition itself is not the main concern of this book, but rather the intense search that this disastrous voyage inspired, a brief introduction is necessary in order to contextualise the material.[8] Since the sixteenth century, repeated efforts had been made by the nations of western Europe to find a northerly sea route for commercial purposes, either a Northeast or a Northwest Passage to Cathay, or China, that was outside Spanish or Portuguese control. The most significant of these was William Edward Parry's voyage of 1819 to 1820 that entered Lancaster Sound in the north-west of Baffin Bay and managed to sail further west than anyone before, as far as Melville Island in Parry Channel, thereby qualifying for an award for sailing past longitude 110°W north of the Arctic Circle.[9] Franklin's expedition was dispatched in 1845 officially in order to complete the Northwest Passage.[10] When no news of his ships was heard for two years, his wife, Jane Franklin (1791–1875), and other figures began to argue that search expeditions should be deployed to offer relief to the missing expedition.[11] Thus began the search for the lost ships that lasted for over a decade and reached its most intensive phase in the early 1850s. Thirty of the thirty-six expeditions were maritime voyages that approached the Arctic by sea; only six were overland or coastal, travelling to the archipelago via the North American coast. The large British naval search expeditions intentionally wintered in the Arctic, using the ship as a headquarters from which to launch sledging journeys in spring that enabled the mapping of thousands of kilometres of what we now know as the Canadian Arctic archipelago in present-day Nunavut and the Inuvialuit Settlement Region.[12]

In the midst of numerous literary studies of polar exploration that include the nineteenth-century Arctic,[13] the role of the visual has been largely overlooked.[14] Much critical scholarly work that exists on the visual culture of the search period traditionally attends more to the public and published imaginings: engravings, lithographs, panoramas.[15] By contrast, little research has been done on the primary visual records of the Franklin search expeditions to the Arctic – the original sketches, watercolours, and drawings. This book fills that gap by prioritising the visual culture of the Arctic during the Franklin search period, both on the ship and in the metropole, without neglecting its textual associations, including inscriptions, captions, and written narratives. In doing so, a broader and more nuanced sense of the cultural history of the Franklin search expeditions is revealed. The sharp focus on a surprisingly neglected time period is essential to do justice to the wealth of archival material that exists from the search expeditions.[16] By drawing our attention to this history, a strong distinction between visual culture on board and that disseminated to the public becomes apparent. While the on-board history has been largely obscured, the metropolitan mode of Arctic representation remains dominant into the twenty-first century.

Interdisciplinary at its core – drawing particularly on art history, the study of literature, and historical geography – the research makes original contributions to knowledge in several fields: the study of nineteenth-century visual culture; travel and exploration literature; and Arctic humanities. As Richard Harding observes, in his appraisal of naval historical scholarship, shipboard life is an aspect that needs more attention; peacetime periods of navies do not attract scholarship in the way that wartime periods do.[17] The Arctic search ships, engaged in a humanitarian cause, can be seen as microcosms of Victorian culture, complete with their own theatricals and illustrated periodicals. Isobel Armstrong notes the nineteenth-century fascination with seeing and observes how visual media bombarded city-dwellers.[18] Allusions to visual technologies (such as the magic lantern and the panorama) permeate published Arctic narratives, showing how embedded these forms of seeing were in the Victorian period. A crucial aspect of Arctic shipboard culture was, as in the metropole, its visuality.

Victorian Visuality and Arctic Exploration

Text and picture are inseparable in the mid-nineteenth-century representation of the Arctic, creating slippage between personal and public perceptions of the region. The visuality of travel in the Victorian period needs to be attended to in conjunction with travel writing. With respect to the latter, Margaret Topping has noted the value of the 'interrogation of images as more dynamic, and potentially contestatory, participants in the narrative'.[19] As Julia Thomas points out, with regard to Victorian illustrated texts, dialogues between words and pictures generate particular stories.[20] Prints and panoramas of the Arctic were further enhanced by their associated texts, informing the viewers and helping to build on the strangeness of the scene in order to convey 'peril' and 'horrors', emphasising the nature of the search for Franklin as a worthy 'masculine' enterprise. Remote locations (such as the Arctic) that most people were never likely to see were alluring and invoked visual curiosity.[21] Kate Flint observes, 'not being able to see with the physical eye is to call into play the powerful forces of imagination and memory'.[22] Glimpses into the Arctic through engravings and lithographs encouraged people to use their own imagination and memory to expand these scenes. Although the high price of large lithographs might seem to indicate that they were seen by a select audience, such media often appeared in printsellers' shop windows or were exhibited at their premises, meaning that they would have reached a far wider audience beyond only those who could afford to buy them.[23]

Although being able to draw was an expected accomplishment in the eighteenth and nineteenth centuries for the upper and middle classes,[24] drawing was embedded in everyday life across ranks and was a normal part of shipboard life. Sketching on the spot encouraged engagement with the environment and was practised by expedition members far beyond any professional obligation. Art historian Geoff Quilley observes that a concern with the visual material from James Cook's three voyages in the Pacific (1768–1780) has eclipsed subsequent visual representation of travel, particularly that created on nineteenth-century expeditions.[25] Cook's expeditions hired professional artists who produced several high art paintings, in contrast to the Franklin search expeditions that relied on the drawing skills of naval officers. Certainly, the rich archive of material from the Franklin search ships has been undervalued and has not benefited from any sustained critical inquiry. This is likely due to the amateur status of the artists, the varying quality of the work, and the attention that the Franklin expedition itself attracts (to the detriment of the comparatively successful search expeditions). Furthermore, work by these amateur artists is more likely to be lost, damaged, or scattered in various archives. Unlike professional artists, who used sketchbooks, many of these works are on loose pieces of paper or within travel journals, making them difficult to discover through searches of online catalogues. With limited resources and challenging environmental conditions, expedition members persisted in creating comic illustrations, travel sketches, coastal profiles, portraits, landscapes, and even theatre backdrops, telling a more complex story than the contemporary media depictions of the Arctic searches.

Small sketches, as opposed to finished paintings, were the most common visual works undertaken in the Arctic by expedition members. Refined commercial products such as prints and panoramas purported to reproduce officers' on-the-spot sketches, thereby associating the contents of their own products with truth and authenticity.[26] By examining a broad array of media, both private and public, I emphasise the significance of these panorama-painters, lithographers, and publishers in re-imagining the Arctic through convincing products. Unlike the Alps, the Arctic was relatively untrodden aesthetic territory, and work that claimed to reproduce original sketches by officers was prized for its 'truth', despite being reinterpreted by professional artists who had a greater aesthetic awareness and less concern with topographical accuracy. Such products of Victorian commercial culture cannot be seen as unmediated perceptions of Arctic 'explorers'. The Arctic had, of course, been represented in paintings that depicted the British whaling industry, and, as the nineteenth century progressed, whaling ships sailed further northwards into Baffin Bay and often got trapped in the ice. In Hull particularly, where the largest whaling fleet was based, artists like John Ward (who it is thought may have been on at least one whaling voyage[27]) and Thomas Binks represented the natural history of the Arctic as well as the whaling ships.[28] However, the focus of these works, in which whale hunting was the main interest, was different from the visual material that emanated from the Franklin searches. The latter was closely connected to the idea of new and ever-expanding horizons, whether real or imaginary, as they sailed further into the unmapped regions of the Canadian Arctic archipelago.

A recurring theme in the critical literature is nineteenth-century British explorers' difficulty in perceiving and representing the Arctic, due to their lack of understanding and unfamiliarity with the environment.[29] Much of the pictorial evidence used to support this argument is comprised of published lithographs, which, it is argued, show that explorers were too concerned with the aesthetic categories of the sublime or the picturesque to be capable of producing an accurate representation.[30] However, although expedition members had an awareness of aesthetic terminology, such an aesthetic concern is not noticeably evident in the primary source material of the Franklin searches. By analysing the transformation from sketch to print and emphasising the importance of the dialogue between picture and text, I argue that it is in the prints and panoramas (products of commercial metropolitan enterprises) that a very deliberate aesthetic manipulation is evident, which intensifies the difficulty of the search.

Much of this manipulation borrowed stylistically from the sublime and the picturesque,[31] aesthetic categories that changed over time, particularly during the eighteenth century when discussion of aesthetics was at its height, with publications such as William Hogarth's *Analysis of Beauty* (1753), John Baillie's *Essay on the Sublime* (1747), and Edmund Burke's *Philosophical Enquiry* (1757 and 1759). According to Burke, the sublime was that which produces delight by the depiction of pain and danger.[32] In nature, the sublime caused 'astonishment' and found its source in obscurity, privation, vastness, infinity, difficulty, magnificence, quick transitions from light to dark, or the idea of physical pain and even the 'angry tones of wild beasts'; indeed, such sources for the sublime were associated with the Arctic in the popular imagination.[33] Absences such as 'vacuity, darkness, solitude and silence', which were a key part of polar imaginaries, were also productive of the sublime.[34] Immanuel Kant's *Critique of Judgement* (1792) maintained that the sublime could only refer to a state of mind, that objects in nature themselves could not be sublime.[35] He divided the aesthetic into the mathematical, an overwhelming of the senses by a huge natural object, and the dynamical, which is an awareness of the effect of power.[36]

Burke's influential *Philosophical Enquiry* had polarised beauty and sublimity, leaving a middle ground that became occupied by Uvedale Price's late eighteenth-century picturesque, which was distinguished by variety, intricacy, and roughness.[37] Price believed that the picturesque was located somewhere between beauty and sublimity and could be blended with either,[38] making it a versatile aesthetic. Thus, dashing waves, Gothic architecture, ruins, and hovels all had picturesque potential. Above all, the picturesque was found in 'partial and uncertain concealment' and in the 'forms, the tints, and the lights and shadows of objects'. However, according to Price, if destruction was threatened, the scene acquired 'a tincture of the sublime'.[39] By the mid-nineteenth century, the picturesque aesthetic had gone through several stages, including being ridiculed in popular satire.[40] These satires made fun of the practice of touring the landscape in search of the ideal viewpoint from which to observe a scene or make a picturesque sketch. Although the formal neoclassical picturesque, as popularised by William Gilpin, divided the painting into the foreground, middle ground, and distance, crucially, picturesque paintings were not expected to be accurate topographical records but to give a general idea of the landscape.[41]

It is important to stress that none of the expedition members in the Franklin searches were professional artists, or even dedicated draughtsmen; they had many other tasks to perform, and personal drawing was only possible when time allowed. Ultimately, although an awareness of aesthetic categories must underlie sketches by amateur artists on the Franklin searches, they were more informed by the environment than they were by the influence of high aesthetics. Members of the Franklin search expeditions who sketched and painted would have also been aware of the strong tradition of marine painting that thrived in nineteenth-century Britain. Although originally influenced by seventeenth-century Dutch marine art, many of the marine artists, like Clarkson Stanfield for example, had experience at sea, and the specialism demanded in-depth knowledge of ships, the sea, and meteorological conditions, all of which members of the Franklin search expeditions would have had. The early romantic interest in marine subject matter, including shipwreck paintings and paintings of the sea, continued during the Victorian period.[42] While some of the on-board work shows the influence of the aesthetic categories (in the choice of subject matter and composition), such drawings underwent considerable transformation by other agents, who often dramatically heightened the sublime, before they reached the public eye. The representation of the nineteenth-century Arctic that we have inherited is the transformed, commercial version of the Arctic, not the version that was represented in the small, self-effacing sketches of expedition members.

Both Robert G. David and Heidi Hansson point out that the Arctic of the popular imagination is associated above all with winter.[43] Hansson further discusses how, in nineteenth-century polar biographies and adventure stories, 'the main adversary is the Arctic itself, almost always in winter'.[44] A common misconception in the twenty-first century is that the Arctic is an empty place that is permanently covered in snow and ice; when the effects of climate change on the Arctic are mentioned, it is not uncommon to hear throwaway responses like 'well, nobody lives there anyway'.[45] Adriana Craciun has observed that a visual and literary focus on ice has dominated the scholarly discourse concerning the Arctic.[46] Certainly, during the Franklin search period, the public image of the Arctic portrayed a space in which a key component of the Arctic narrative – 'man's struggle against the elements'[47] – could be displayed. The imaginary of an empty Arctic has been associated specifically with (published) representations by nineteenth-century British explorers. Francis Spufford suggests that 'when it comes to the explorers' success or failure at traversing the landscape', Inuit 'did not belong in the stories of discovery and achievement'.[48] For David, the 'single-minded objective' of the mid-century Franklin search expeditions resulted in a sudden disappearance of Indigenous people from representations in favour of subjects that depicted exploration.[49]

While it is true that a considerable proportion of the published pictures and texts from the search period do not show an Indigenous presence, one reason for this is that many of the large, high-profile search expeditions of the 1850s spent the majority of their time in the northern part of the Canadian archipelago (as seen in Figure 0.1), an area that did not support Indigenous populations in the mid-nineteenth century.[50] Of course, prior to European incursions, there had long been a human presence in some parts of the Arctic,[51] and in the mid-nineteenth century, the far north-east of Siberia on the west side of the Bering Strait was inhabited by the Chukchi and Yupik; Iñupiat and Yup'ik were present in coastal north Alaska; and Inuvialuit and Inuit were scattered over the coastal High Arctic of Canada and in Greenland.[52] However, a large part of the central and northern half of the Canadian archipelago was uninhabited during the period of the Franklin searches. When ships did spend significant amounts of time near settlements, particularly in the Bering Strait region, there is ample literary and visual material in the archive attesting to a peopled Arctic. Intense interaction took place when ships wintered near Indigenous communities, and expeditions were reliant on the goodwill of residents for their hunting skills, winter clothing, and geographical knowledge, as well

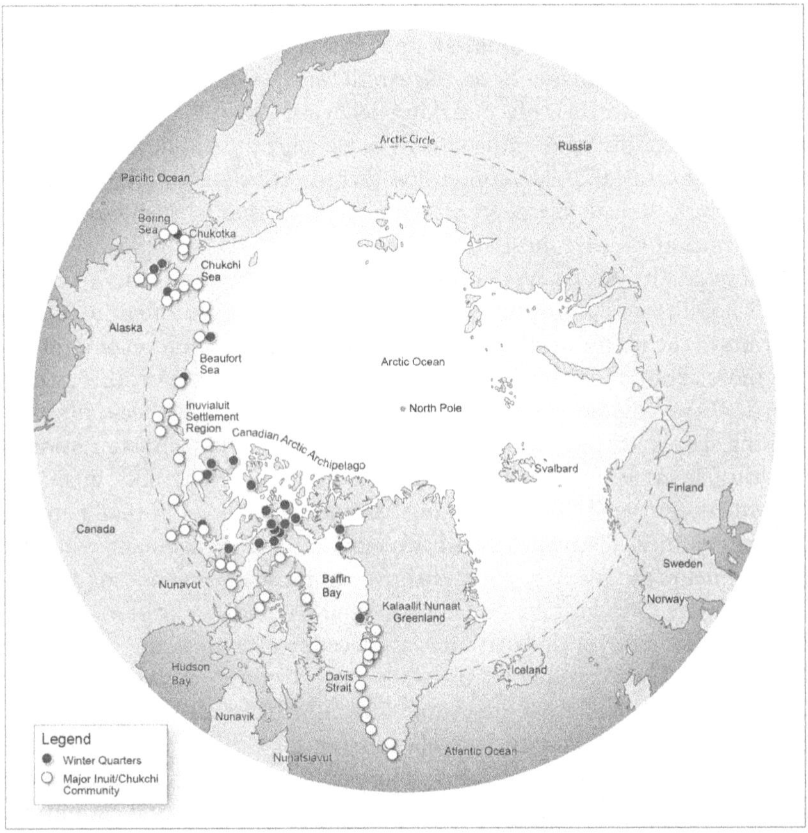

Figure 0.1 Map of the Arctic showing winter quarters of the maritime search expeditions and Indigenous communities in areas closest to expedition routes and winter quarters, 1848–59. Map by author.

as social, and sometimes sexual, interaction. There are also published versions of this peopled Arctic. For example, a large portion of the narrative *Ten Months among the Tents of the Tuski* (1853) focuses on the expedition's interactions with Chukchi individuals.[53] As Efram Sera-Shriar has observed, reliable travel narratives contained substantial evidence that Indigenous knowledge and social interaction were valued,[54] even if this value was not fully acknowledged.

The archival, and indeed some of the published material, shows evidence that complicates the notion of Mary Louise Pratt's imperial gaze. This gaze sees the landscape in scientific, objective terms, observed by 'the

European male subject of European landscape discourse – he whose imperial eyes passively look out and possess'.[55] But the relationship of the search expedition members to the Arctic and its people was far more complex than that; their place-making behaviours as well as their interpersonal relationships with Indigenous people suggest a multifaceted gaze, which supports Majid Yar's contention that the association of the gaze with power overlooks other aspects of vision such as the 'hermeneutic, emotional, communicative' possibilities.[56]

Although the Arctic was considered to be an environment in which man's nobility was put to the 'ultimate test',[57] the establishment of winter quarters in the ice meant that life became less about exploration and travel, and more about home-making. Arctic winters in particular became associated with specific activities – masked balls, pantomimes, ice-sculpting, evening classes – thus creating a sense of history, identity, and community, resulting in a feeling of belonging and home.[58] Expedition members individually engaged in place-making behaviours, such as informal naming of topographical features,[59] social interaction, and engagement with the environment; the primary sources show significant *topophilia*, or attachment to place.[60] Indeed, based on John Agnew's triadic definition of place, the Arctic took on all the attributes of place for overwintering expedition members.[61] These visitors to the Arctic, who ceased to move through a space, began to behave a little more like inhabitants and less like Pratt's British explorer, or 'monarch-of-all-I-survey', in Africa.[62] Neither were overwintering expedition members tourists in search of the picturesque, moving through a landscape; they were not perpetually apprehending the sublime that travellers first began to experience as they crossed the Alps from the late seventeenth century. They were living and working, embedded in the environment and forming habits of familiarity through daily routine. Both the sublime and the picturesque are dependent on novelty and on the visibility of new prospects. In contrast, the overwintering maritime expeditions were immobile for much of the year, and some expedition members made repeated voyages to the Arctic, even returning to the same places several times.

The imagined geography of the Arctic, as a vast empty space where man pits himself against a hostile environment, has been created and maintained by public visual representations and texts.[63] The idea of North is projected as a masculine gendered zone.[64] Nineteenth-century Arctic exploration is typically represented in terms of hardship, pain, and suffering; titles of books such as *Arctic Hell-Ship* (2007) emphasise this aspect for potential readers.[65] Indeed, very little has been written about the positive aspects of polar exploration.[66] However, close examination of on-board visual and written records yields many positive responses to the Arctic and reveals an expectation of future nostalgia for the Arctic among members of search vessels, even those who participated in particularly gruelling expeditions,[67] reminding us that, as Lawrence A. Palinkas and Peter Suedfeld note, 'positive and negative reactions' to polar conditions are not mutually exclusive.[68] This book reveals overlooked details in the archive – a smiling face in an ostensibly unpleasant situation, a revealing phrase in a manuscript – that complicate the notion of a masculine space of peril and danger and highlight a different way of imagining the Arctic. According to the president of the Royal Geographical Society in the late 1850s, Roderick Murchison, the searches for Franklin had proven to be 'in times of peace . . . the best school for testing, by the severest trials, the skill and endurance of many a brave seaman'.[69] In fact, these trials of masculinity also involved 'unmasculine' chores like doing the laundry and cooking, as well as the enthusiastically pursued pastimes of dress-making and cross-dressing for on-board theatre performances.[70]

When we explore the archives behind the attractive public pictures, a far more complex visual matrix is observed than the immense icebergs and ships in jeopardy that dominate the public representations. Even today, nineteenth-century prints lavishly illustrate twenty-first-century texts and continue to inform our ways of thinking about the Arctic. As Julie F. Codell points out, visual culture is often marginalised, 'or worse, is used to illustrate' history, colonialism, and historical events.[71]

The lithograph *Critical Position of H.M.S. Investigator* (Figure 0.2), 'the most arresting of all polar images',[72] is frequently presented and consumed as a factual artefact (showing that the aura of authenticity still persists), yet the scene of peril is the work of a London-based lithographer.[73] Despite this, *Critical Position of H.M.S. Investigator* has endured to become the visual catchphrase of all expeditions that searched the Arctic in the mid-nineteenth century.[74] Although contemporary reviewers of prints were at pains to point out that certain published 'sketches' were 'facsimiles' of those by officers, their very need to state this fact betrays a distrust of popular visual culture as a source of information; their concern with the 'truth', 'accuracy', and 'reality' of representations paradoxically shows their awareness of the power of the visual as a tool of deception.

Visual Culture and Arctic Voyages is led by the visual material itself, acknowledging the immediacy, availability, and proliferation of images in the Victorian period. My analysis of sketches, illustrated periodicals, published texts, panoramas, and prints blends methodologies from the

Figure 0.2 Samuel Gurney Cresswell, *Critical Position of H.M.S. Investigator on the North Coast of Baring Island, August 20th 1851*, 1854. Lithograph, 44.3 × 61.2 cm. Courtesy of Toronto Public Library.

disciplines and fields of publishing and book history, the study of literature, art history, visual studies, historical geography, the history of science, and cultural anthropology. In particular, the interpretative approach of cultural history, which acknowledges the creativity and subjectivity of the human mind while also embracing a holistic view identifying overall structures or patterns,[75] informs the research methodology. Close reading, cultural analysis, and semiotic analysis all have a role to play in the interpretation of the material. In common with Susan Close, who considers photography to be a 'social practice',[76] visual culture on the Franklin search expeditions can be thought of as inherently social, not only engaging with subjects but connecting practitioners to each other through their visual work.

Art historical critical skills that account for the materiality of images as objects and attention to subject matter and social and art historical contexts are brought to bear on a very broad range of visual media encompassed by visual culture studies. The concept of visuality 'attempts to grasp the visual aspects of the relationality and performativity of human life (or of subjects) in societies and cultures'.[77] Critically examining the visual archive means taking into account the role of audiences, contexts of production, and display.[78] Visiting archives and libraries enabled the discovery of uncatalogued pictures, the examination of detail, the observation of condition, close-up photography, and the recording of core information.[79] Such depth of observation and recording allows critical engagement with the representations and an appreciation of the richness of the material.

Over the course of my research, it became apparent that, while some critical scholarship uses pictures from the Franklin searches to illustrate publications, there is little formal attempt to understand and analyse these pictures, to bring any depth and meaning to their contents. Popular illustrated histories use pictures to extend their text, decorate, and give visual pleasure.[80] This book is informed by the archival nature of the research, beginning with the pictures and texts themselves, rather than 'starting from a series of observations and assumptions imposed on a body of material'.[81] Like photographs, sketches and their captions or associated texts can be seen as 'visual incisions through time and space',[82] that 'spring leaks' on interrogation of their 'distinctions and points of fracture'.[83] Their interaction often reveals a discord between the written and the visual. Historical geographer Felix Driver stresses that published narratives of exploration merely represent the ultimate stage in a chain of stages.[84] While this project includes published exploration narratives in its dataset, their analysis is embedded in an awareness of the archival sources and contexts that led to their creation. Driver notes that such published texts cannot be seen as single products of imperialism, but that they would be better viewed as 'articulations of practices',[85] having gone through many stages. The same is true of published prints, which have gone through a similar process of alteration and are preceded by layers of archival material. Furthermore, as Bernard Smith explains, we can think of 'three principal means by which drawings may be said to represent, we might call them inventive drawing, illustrative drawing and documentary drawing'. They are based respectively on images from the mind, things that are described in words, and 'things that the draughtsman perceives out there in the world'. These categories are not well defined, and he invites us to think of all three means as 'potential components of perception'.[86]

The importance of the visual material to this book necessitates an explanation of the hierarchy of descriptors that will be used throughout.

Firstly, the term *picture* is a general (primary) term for all two-dimensional representations (including prints, paintings, sketches, drawings, photographs, and panoramas). As Leila Koivunen suggests, *picture* can be more useful than *image* as it distinguishes the material from literary images.[87] *Sketches* are generally rough drawings, made quickly in ink or pencil; the term *drawing* implies a more finished sketch. The term *illustration* can be used when a picture was created specifically to complement printed text and is influenced by that text. A *print* is 'formed by transfer from one surface or source to another. Usually created with ink(s) and produced in multiple impressions.'[88] *Relief printing*, such as woodcuts, is the oldest known technique, whereby non-printing surfaces are cut away, leaving raised surfaces to hold the ink. The ink is then transferred to paper through pressure.[89] *Intaglio* methods, like *engraving*, *etching*, and *mezzotint*, involve incising a plate, the recessed grooves of which hold the ink. Much greater pressure is then used to transfer the ink to paper.[90] *Planographic* or surface printing is the most recent technique and employs a flat printing surface. *Lithography* is the original planographic technique, whereby the picture is 'printed from a stone or metal surface on the principle of antipathy between grease and water'.[91] The term *printer* includes *engravers* and *lithographers* who reproduced *pictures* for publication. The nineteenth century saw the rise of many new and evolving forms of visual entertainment. The original *panorama* was a large 360-degree painting that encircled the viewer and, by eliminating external reference points, immersed the viewer in the scene. The *moving panorama* consisted of a long canvas that was unrolled past an audience, often accompanied by music. The nineteenth-century *diorama* was a 'large, partially translucent scenic painting, which by means of varied illumination simulated such effects as sunrise, changing weather, etc.'.[92] A *tableau* was an arranged scene of objects or people, and a *magic lantern* was a type of image projector that became increasingly popular in the mid-nineteenth century.

Central to this book is a conviction that we cannot interrogate the literary or visual material, much of which concerns the environment, without acknowledging the complex and varied physical geography of the regions in which the Franklin search expeditions found themselves. A considerable portion of the archival material I will discuss was produced in the Arctic – a place that differed vastly from metropolitan production contexts. To this end, underlying the research is a knowledge of nineteenth-century and modern maps and an attentiveness to regional climatic and meteorological variation.

Geographical and Historical Background

The Arctic region is an immense area not easily delineated and with complex connections to lower latitudes.[93] The Arctic Circle, at a latitude of approximately 66° North, is often defined as the southern boundary of the Arctic (Figure 0.1).[94] This marks the latitude at which the sun remains above the horizon at the summer solstice and below the horizon at the winter solstice. Other definitions use the zone of continuous permafrost on land and the extent of sea-ice[95] or refer to the Arctic as the region north of the +10°C July isotherm (a line that roughly equates to the northern treeline and includes areas south of the Arctic Circle), where the average temperature does not exceed +10°C in July.

It is vital to understand that the Arctic is not one uniform region; the environment is varied and the conditions local. For example, in the 1950s, Inuit families from Inukjuak, Nunavik (northern Québec) that were forcibly relocated almost 2,000 kilometres north to Resolute Bay (in the region where so many of the search expeditions wintered) found 'their traditional knowledge and hunting techniques were out of place' and the 'total darkness from November to February was unfamiliar and disabling'.[96] Tellingly, one of the Inuktitut names for Resolute, Qausuittuq, translates as 'the place with no dawn'.[97] It is worth reiterating that many of the large search expeditions wintered to the north of Parry Channel, in areas that Inuit themselves did not inhabit.[98] The Canadian Arctic archipelago of Nunavut and the Inuvialuit Settlement Region reduces in elevation from east to west, from a height of over 1,500 metres on Devon and Baffin Islands to average elevations of 300 metres on Banks Island in the west. Expeditions that entered via the Pacific Ocean and the Bering Strait encountered the mountainous region of the Chukchi Peninsula in north-east Siberia, the coastal plain of the Alaskan Arctic, and the west side of the archipelago, including low plains and plateaus.[99] In the mid-nineteenth century, the national and provincial borders of the Arctic that we are familiar with today largely did not exist. Alaska was part of Russian America, and the remainder of the northern part of the continent was known as British North America. Much of the archipelago was unknown to Europeans. The north and west of the present-day Canadian mainland were dominated by the Hudson's Bay Company, which held a fur-trading monopoly controlled from London.[100] The Province of Canada was the largest colony in British North America, where the settlements of Ottawa, York, Québec, and Montréal were located.[101]

Franklin was no stranger to expeditions in the Arctic, having participated in, or led, three expeditions between 1818 and 1827. For the 1845 expedition, his ships had been provisioned for a voyage of three years, and it was not until 1848 that the first major maritime expedition was sent by the Admiralty to search for the missing vessels. By August 1850, eleven ships were in the region of Lancaster Sound when Franklin's first winter quarters, including the graves of three seamen who had died from tuberculosis and pneumonia, were found on Beechey Island.[102] John Rae returned from an overland expedition in 1854 with evidence of the Franklin expedition's fate – in the form of 'relics' and Inuit accounts that suggested the starving crew had resorted to cannibalism.[103] The Admiralty ceased its search and turned its attention to the Crimean War (1853–6), but there was a desire among other agents, such as Jane Franklin, who campaigned for continued searches, to be provided with more evidence for what exactly had happened. Hence small-scale searches continued to be carried out until Francis McClintock returned in 1859 with additional first-hand evidence including objects, a written note, and reports of skeletons near King William Island.[104] By this stage, the main elements of what had happened were known. As Loomis succinctly notes: 'Franklin had died early in the expedition; his ships had been either crushed by the ice or deserted by their crews; the men had died one by one as they tried to walk south to the Canadian mainland.'[105]

By far the largest and longest searches were the British naval maritime search expeditions. Between 1848 and 1852, six of these were dispatched, with the last expedition to return to Britain doing so in 1855 (see Table 1).[106] Additional smaller naval expeditions served as supply and relief vessels to the main search expeditions, and several private expeditions, funded or inspired by Jane Franklin, were also involved. Indeed, Behrisch Elce credits the tireless letter-writing campaign of Jane Franklin with the perseverance of the search and argues that her influence led to the 'opening up of one of the world's most mysterious places'.[107]

All of the six large naval search expeditions spent a minimum of one winter in the Arctic; the *Plover* spent six years in the Bering Strait region, with some of her crew volunteering to remain aboard for that entire length of time.[108] Other expedition members volunteered to return on subsequent voyages, increasing their familiarity with the region and developing the particular culture of Arctic maritime expeditions. The search ships aimed to enter the archipelago in summer and find a sheltered bay in which to winter over. As the sea iced over,

Table 1 *Details of the British naval maritime expeditions that searched for Franklin. Other naval expeditions functioned as supply and rescue endeavours to these expeditions already in the Arctic.*

Year	Expedition Leader	Ships (In Command)	Approach
1848–9	James Clark Ross	*Enterprise* (James Clark Ross) *Investigator* (Edward Bird)	Atlantic
1848–54	Thomas Moore Rochfort Maguire	*Plover* (Thomas Moore 1848–52; Rochfort Maguire 1852–4)	Pacific
1850–1	Horatio Austin	*Resolute* (Horatio Austin) *Assistance* (Erasmus Ommanney) *Intrepid* (John Cator) *Pioneer* (Sherard Osborn)	Atlantic
1850–4	Robert McClure	*Investigator* (Robert McClure)	Pacific
1850–5	Richard Collinson	*Enterprise* (Richard Collinson)	Pacific
1852–4	Edward Belcher	*Assistance* (Edward Belcher) *Resolute* (Henry Kellett) *Pioneer* (Sherard Osborn) *Intrepid* (Francis McClintock) *North Star* (William Pullen)	Atlantic

preparations were made by roofing the top deck with sailcloth, building insulating snow walls, and lighting the stoves. The three darkest months were from November to February, but even then, the polar twilight made it possible to read a book outside at noon on the shortest day of the year. During the spring, exploratory sledge journeys were undertaken that could cover hundreds of miles over the ice. If returning to Britain, ships aimed to exit the archipelago in late summer, when the ice broke up, and could arrive back to the metropole by autumn, when the press announced their return.

The structure of this book is informed by media type and by the geographical and chronological specificities of provenance. The first two chapters examine the type of visual material created on ships in the Arctic, and the final three chapters focus on visual media in the metropole. This provides a better understanding of how the Arctic was transformed (how the contents of on-board material were selected and altered to create prints, panoramas, and texts), with the latter chapters informed by a knowledge of the former. On occasion, the use of case studies enables close examination of individual works and reveals processes of transformation.

Chapter 1, 'On the Spot', investigates the practice of drawing and painting aboard ship in the Arctic during the Franklin searches, revealing how sketches were made for various reasons, many of them unrelated to the more obvious purposes of science or navigation, and showing how close attention to their attributes allows new layers of information to emerge. The second chapter, 'Breathing Time', further examines the visual culture aboard ship through an exploration of illustrated on-board periodicals, which were a key part of the maritime culture during the Franklin search. These human-centred productions turn inwards to observe the ship's inhabitants in winter quarters, focusing on social interaction and incidents. The Arctic itself and expedition members' incongruous domestic life was the source of a humour that was personal and particular to the expedition members' situation.

Moving then from the ship to the shore, I examine how the Arctic was presented to a public metropolitan audience and, where possible, show the on-board raw material from which these public pictures developed. Chapter 3, 'These Dread Shores', explores the text–picture interplay in narratives of travel and exploration that were published by members of search expeditions, revealing tensions between the domesticated and the sublime Arctic. In the fourth chapter, 'Never to be Forgotten', the practice of creating large-format paintings known as panoramas is examined in relation to the search, showing how a metropolitan audience anticipated education, entertainment, and an aesthetic experience from Robert Burford's *Summer and Winter Views of the Polar Regions* (1850) at Leicester Square, London. Finally, Chapter 5, 'Power and Truth', observes how the Arctic and the act of searching were represented in folios of lithographs produced from officers' sketches.[109] By privileging the visual, the lithographs provided audiences with very specific Arctic narratives.

Visual Culture and Arctic Voyages shows how the appropriation of the story of Franklin, and the search parties that followed, into popular culture has evolved from the first-hand records of expedition members who went to the Arctic. Yet these records were transformed into commodities before they reached the public eye, often concealing the version of the Arctic that was local, intimate, and familiar. This local Arctic was fertile ground for the production of culture, with expedition members from all ranks partaking in representation. Moreover, the Arctic was exploited for humorous ends, with expedition members mining the incongruous nature of their situation. The narratives, prints, and panoramas that reached the metropolitan audience, however, displayed an icy, threatening world, which still reverberates in the popular imagination today.

While Jane Franklin, perhaps, searched for some clue as to the whereabouts of her husband at Leicester Square on that Saturday in early February 1850, William Browne (who is likely to have been present among the 'gentlemen who had wintered in these northern latitudes'[110]) must have marvelled at how the creators of the panorama had transformed his on-the-spot sketches into a dramatic Arctic spectacular. This book is about such transformations, and how the increasingly familiar Arctic became alien and strange once it was transported to the metropole. In order to understand the visual complexities of these representations, it is necessary to begin by travelling to their source, by discovering the shipboard visual culture of the expeditions that sailed northwards in search of John Franklin and his men.

CHAPTER 1

'On the Spot'
Scientific and Personal Visual Records (1848–1854)

In the summer of 1848, as the expedition led by James Clark Ross (consisting of the ships *Enterprise* and *Investigator*) made its way slowly northwards along the west coast of Greenland, the young lieutenant William Browne (1821–71) must have been busy sketching aboard the *Enterprise*. He would, by now, have had a strong awareness of his abilities as an artist. Already his drawings from naval surveying in the Pacific had been engraved for Edward Belcher's *Narrative of the Voyage of H. M. S. Samarang* published that same year.[1] It is curious, then, that only three works from Browne's first expedition to the Arctic are to be easily found in public archives. The sketches with which he returned generated not only Burford's large panorama but also a folio of ten lithographs and a set of engravings in the *Illustrated London News*.[2]

As further proof of his productivity and enthusiasm, we might look to one of the three works that remains: *Rough Sketch Moored to an Iceberg* (Figure 1.1) is a small work on grey paper that shows two ships in the ice and has three figures in the foreground.[3] At first glance, the picture is unimpressive and easily dismissed. However, attending to the details reveals valuable information. At the top of the page, along with the explanatory title, 'Willy Browne (Lieut) no. 37' has been hastily written, implying that at this early point in the journey Browne had already produced over thirty drawings. Certainly, this note suggests that many more of his works could be recovered in the future, or that tens of small and apparently insignificant sketches have been discarded over time. We might also observe the very small size of this work (10 × 19 cm) and the limited colour palette of blue, white, black, and brown. Looking again, we notice the stains on the paper, the obvious speed with which the drawing has been executed, and the use of crayon as a practical choice for outdoor drawing. In short, the sketch tells us far more than simply what it illustrates. It contains obvious clues to its origins on the spot, outdoors, in the Arctic, and textual information that hints at the amount of visual material from these expeditions that must

Figure 1.1 William Henry Browne, *Rough Sketch Moored to an Iceberg*, 1848. Crayon, 10.2 × 18.8 cm. Scott Polar Research Institute, University of Cambridge, with permission.

have been lost, destroyed, or still lies undiscovered in the attics of Victorian homes.

Such visual archives that remain are often small and fragile, created on paper with pencil, watercolour, crayon, or ink, making them vulnerable to fading, damage, and loss. What remains today can be found catalogued in archives, hidden between the leaves of journals, or in substantial sketchbooks. The works from the Franklin search expeditions are, without exception, by amateur artists, who often drew in addition to their duties on board ship, using minimal, rudimentary materials. The works are frequently unsigned, indicating that their makers had no pretensions towards 'art'.[4]

Considering the attention given to Franklin's final Arctic expedition by present-day media, popular authors, and academic critics, very little scholarly research has been done on the primary visual records of the search expeditions that followed in Franklin's wake. Although academic work has discussed the Arctic imaginary that evolved from this fateful voyage, what has been done on the visual culture attends chiefly to the public face of the Arctic in the mid-nineteenth century: the engravings, lithographs, exhibitions, and professional artworks.[5] These public representations are often quite different in style and content from the hundreds of unpublished documentary drawings and paintings by expedition members that exist in archives scattered across the world. The unpublished visual records have, in general, not been studied to any great extent by art historians, who traditionally focus on the more established canon, and both historians and literature scholars are, understandably, more attentive to the written record.

In an article on nineteenth-century naval topographical drawings in the Pacific from an art historical vantage point, Geoff Quilley notes that his own study was 'richly suggestive of the potential significance of the wealth of similar material that lies, generally disregarded or overlooked, in archives and repositories around the world, and which now needs scholarly attention'.[6] The visual records of the Arctic search expeditions are also significant in the way in which they capture responses to the environment and how they reveal a different Arctic to that which was presented to the public. More recently, Felix Driver has noted that, in the context of the history of geography and exploration, the 'practice of drawing has received some attention though arguably rather less than it deserves'.[7]

All officers on naval expeditions, midshipmen in particular, were expected to make charts and coastal profiles for navigational purposes: 'the ability to render in graphic form the dimensions, detail and colour of coastal landscapes was . . . an essential aspect of the surveyor's task'.[8] They required the skills, techniques, and materials to produce accurate topographical drawings, sometimes gaining the knowledge by training at the naval college in Greenwich but often acquiring the skill set aboard ship. Ships' surgeons too were expected to be able to record visual information as part of their contribution to the study of natural history. As Nanna Katrine Lüders Kaalund notes, the Franklin search allowed for the development of '"opportunistic" science' in the Arctic.[9] However, science and navigation were far from being the only motivations behind the pictures drawn by members of expeditions. In many instances, men were clearly drawing for their own personal reasons rather than by any compulsion to fulfil their duties aboard a vessel. Neither were officers the only ones who drew and painted; evidence exists in the archive that visual recording was practised among all ranks of a ship's company.

This chapter aims to uncover some of the richness of the on-board visual records, while showing what further information can be gleaned from these lesser-known works. Officers who created drawings to advance knowledge attempted to give 'objective' renderings of what they observed, yet these pictures could have subjective overtones, and on occasion there were underlying expectations that they would be aesthetically pleasing. However, in addition to the creation of scientific drawings for geographical and navigational purposes, the desire to have personal visual souvenirs to serve as nostalgic reminders of places, and even people, encountered on a voyage provided an equally strong impulse to draw. Some of these drawings eschewed any scientific intent, aligning themselves rather with subjective experiences in the Arctic, and yet could still inadvertently contain scientific information. In this way, the objective and subjective permeate each other and can be seen to merge in some of the works. In many cases, inscriptions and other forms of paratext (inscribed frames, for example), or the lack thereof, reveal the artist's intentions and the projected future purpose of the record. Furthermore, the primary visual corpus examined confirms that summer views (when on-the-spot work was most feasible) dominate the archive, contrary to the popular imaginary of a region covered in ice and snow.

Although the argument that nineteenth-century British explorers in the Arctic were overly influenced by the aesthetic categories of the picturesque and the sublime is a popular one,[10] an examination of the archival material does not show that the picturesque or the sublime dominated representations of the region, although expedition members probably had an awareness of such categories. Significantly, we should also note that, in the library of the *Assistance* during the Belcher search expedition from 1852 to 1854, there were no books by William Gilpin, Edmund Burke, Uvedale Price, or any other writers on aesthetics, out of a collection of almost seven hundred volumes, in which science, geography, and history dominated.[11]

The archive of expedition members' visual records is scattered; from Canada to Britain to Australia, the dispersal of the records suggests that not all material was handed over to the Admiralty on the expeditions' return or that some material was returned to expedition members.[12] Some of the visual material existed in private collections and subsequently found its way to local and national archives; it is conceivable that still more lies buried in family homes worldwide. Thus, we must be mindful that the work viewed during the course of this research represents only a portion of an incomplete and fragmented visual register. The fragile nature of the works by these amateur artists implies that much more material may simply have been disposed of; such small, unsigned works, faded and possibly torn, may appear unimportant and do not lend themselves well to survival. Even so, this study found over five hundred sketches, drawings, and paintings relatively accessible in archival repositories. In addition, some early photographs survive and are included in the discussion. Only a small portion of the known archive can receive attention here. I have chosen examples by expedition members of all ranks and examined the rhetoric of these pictures to show the nature of the visual culture aboard ship. Some of these pictures are typical of naval topographical drawing, others are unusual, all show that the culture of visual representation aboard ship was powerful and that an examination of these pictures yields a more nuanced view of the life aboard Arctic search vessels. Throughout the book, I refer to similar primary documentary records to make my argument when discussing the visual representation of the Arctic in the mid-nineteenth century.

Drawing as a Scientific Instrument

Hydrography, or making maps for maritime navigation, was a key science on naval expeditions, and technical expertise as a representational artist was valued. Important to navigation in particular was the accurate visual recording of coastlines. Coastal elevations could display geological information if the stratigraphy of the land was accurately drawn.[13] However, from the late eighteenth century, the British Admiralty ceased to use professional artists on survey and exploration voyages.[14] Hence, the many expeditions that searched for Franklin contained no professional artists; rather, officers, with varying levels of talent, experience, and technical knowledge, were expected to draw for navigational purposes. Such variations are seen in individuals' choice of materials and methods, handling of perspective, confidence of line or brushstroke, and in their ability to draw human and animal figures (understandably, many naval officers were far more adept at drawing ships and displayed hesitancy in depicting human figures). As Felix Driver points out, drawing was a way in which a midshipman or officer could distinguish himself and perhaps gain a promotion. Carefully produced drawings could impress superiors at a time when there was intense competition for naval posts.[15] Notwithstanding the possibility of a promotion, many on board were drawing for other, less career-motivated, reasons – to create mementos of their experiences, to pass time during the winter, or to record their subjective response to the environment.

The library catalogue printed on board the *Assistance* in 1853, during the Belcher expedition from 1852 to 1854, lists only two titles on painting techniques ('Painting Water Colors' and 'Painters Art. Fields, Rudiments of')[16] out of an estimated collection of 680 books.[17] There were also two books on photography, reflecting an interest in adopting the new techniques of visual recording ('Photography' and 'Photography Thornwaits [*sic*], Guide to').[18] The largest portion of the library, over a quarter, consisted of works pertaining to Geography (including travel narratives that contained illustrations).[19] Although there were no books listed that dealt explicitly with the aesthetics of the sublime or the picturesque, the library on the *Assistance* did contain volumes of Alexander von Humboldt's recently translated *Cosmos*, which included his essay 'Landscape Painting'.[20] Humboldt's essay asserted the link between science, art, and poetry, arguing that 'an ancient and deeply-seated bond' united 'natural knowledge with poetry and with artistic feeling'.[21] Science was closely tied to landscape painting in its representation of the 'characteristic physiognomy of different positions of the earth's surface'.[22] Nigel Leask comments that Humboldt regarded 'aesthetic and emotional responses to natural phenomena' as additional data. Although Leask draws a contrast between this approach and the 'rigorous exclusion' of those responses 'from contemporary practices of naval and military surveying',[23] certainly during the Franklin search expeditions, personal responses sometimes permeated the officers' scientific work that was intended to add to knowledge. Skills acquired for scientific and navigational recording could also be put to personal use, and the two modes of expression often overlap in the archive, as more private records can still retain scientific information.

Although the primary object of the searches was to find the Franklin expedition and render assistance, the officers on the voyages were expected to accumulate information for the advancement of science. Trevor Levere notes that the principal arguments for the exploration of the Arctic at this time were scientific rather than commercial, and Andrew Lambert draws attention to the hugely influential role of Edward Sabine (a founder member of the Royal Geographical Society (RGS) and later General Secretary of the British Association for the Advancement of Science (BAAS)) in encouraging expeditions searching for Franklin to collect scientific data, particularly in relation to geomagnetism and navigation.[24] Science became increasingly visual during the nineteenth century, as prints were more available to the public and spectacles like the panorama, which sought to give 'an intoxicating God's eye view' of a scene to observers, were popular forms of entertainment as well as education.[25] A 'craving for visual images' was a characteristic of mass culture in the latter half of the nineteenth century.[26]

Although topographical draughtsmen were not in the same class as professional artists,[27] the painting of landscapes became both popular and diverse in Britain during the late eighteenth and early nineteenth centuries. As Luciana Martins notes, this period saw both the classical ideals of landscape and picturesque taste confronted by a new naturalism in landscape painting.[28] This new naturalism was accompanied by an interest in natural science and in understanding nature.[29] Furthermore, landscape, and the artist's response to it, was seen to be capable of representing ideas and events, emotions and individuality. The idea of fusing scientific observation with subjective feeling was a combination that became influential in the development of landscape painting during the nineteenth century.[30] The boundaries between aesthetics and natural science became more fluid in practices of visual representation.[31]

The Admiralty Manual of Scientific Enquiry was published in 1849, as the Franklin searches were gaining momentum, and it gave officers a summary of expectations in the various fields of science; Erika Behrisch Elce notes that 429 copies of the first edition went directly to the Admiralty.[32] The library catalogue of the *Assistance* in 1853 listed three copies of the manual.[33] It was intended specifically for the use of 'Officers in Her Majesty's Navy' as well as for 'Travellers in General' and was designed to provide basic guidelines in the natural sciences so that travellers could collect information useful to science.[34] Naval officers were expected to take accurate observations and report their findings but were not expected to produce theoretical findings.[35] The Memorandum at the beginning of the text states the Admiralty's aims in commissioning the book:

> It is the opinion of the Lords Commissioners of the Admiralty that it would be to the honour and advantage of the Navy, and conduce to the general interests of Science, if new facilities and encouragement were given to the collection of information upon scientific subjects by the officers, and more particularly by the medical officers, of Her Majesty's Navy, when upon foreign service.[36]

Chapter by chapter, it deals with Astronomy, Magnetism, Hydrography, Tides, Geography, Geology, Earthquakes, Mineralogy, Meteorology, Atmospheric Waves, Zoology, Botany, Ethnology (an emergent discipline at the time), Medicine, and Statistics. The ability to be an accurate draughtsperson – something that required time, experience, acute observation skills, and concentration – was tied in closely with the practice of scientific observation and investigation. Although it did not appear to warrant its own chapter, draughtsmanship is referred to repeatedly throughout the manual. Drawing, both as an interpretational tool and as part of the recording process, is recommended over a lengthy verbal description. Charles Darwin, who wrote the chapter on Geology, recommended a book by the natural historian Louis Agassiz on glaciers, 'with its admirable plates', to anyone going to the polar regions.[37] The plates in Agassiz's *Études sur les Glaciers* (1840), drawn by Agassiz's artists and printed on his own lithographic press, are detailed and scientific, yet also display an artistic understanding of composition and a delicate treatment of line, showing how geology drew art and science together in its depiction of the environment.[38] The implication here is that other artists should aspire to depict the landscape as elegantly, clearly, and accurately as in those plates. The chapter on Geography by William J. Hamilton recommends that

> slight sketches of the country, and of the peculiar forms of hills, however hastily and roughly made, will be of more assistance in recalling to his own mind, or in making intelligible to others, the features of the district he has visited, than long and elaborate descriptions. Let him then acquire the habit of never quitting his ship without his note-book and pencil and his pocket-compass.[39]

Here, Hamilton recognises the power of memory triggered by the sight of a sketch that assists in the recollection of topography. Such sketches, by their making, inscribe features of the landscape on the mind of the maker. The role of science and geography in encouraging engagement with the landscape and providing the skills and materials for visual expression is important. However, expedition members' interest in and knowledge of science in general must have varied greatly. Regarding the experiments in magnetic variation on the *Investigator* in 1848, Samuel Gurney Cresswell, appointed as a mate, wrote to his parents: 'The scientific men have been making no end of observations, a whole lot of dodges to find the dip of the needle &c. I not being a scientificer do not understand much about it.'[40]

The scientific rigour with which some expedition members approached their work is evident in the many drawings that exist in the Admiralty archives. In the case of the clerk James J. Rutter, aboard the *North Star* from 1849 to 1850, his visual work shows that he too aspired to representational accuracy even though drawing was not strictly part of his duties. The Admiralty instructions regarding visual records given to Master James Saunders on the *North Star*, which was intended to re-supply the Ross expedition of 1848, indicate the information required from topographical sketches of the 'adjacent shore', which needed bearings and the 'latitude of some point … determined, when practicable, by artificial horizon, on shore'. Longitude needed to be established by 'careful chronometric observations, or equal altitudes'.[41]

The master's assistant, John F. R. Aylen, had the responsibility of drawing maps and coastal profiles, but Rutter seems to have taken on the task of drawing 'views' of the surrounding environment. All of the surviving work by Rutter and Aylen is dated between June and September, strongly suggesting their work was undertaken outdoors and on the spot. The *North Star* was unable to locate either of the Ross ships and spent an unplanned winter in Wolstenholme Sound, north-west Greenland from October 1849 to July 1850.[42] *A View of the Three Glaciers at the Eastern Extremity of Wolstenholme Sound* is torn and inscribed with the words 'sketched and drawn from nature'.[43]

The watercolour painting, measuring 22.2 × 35 cm, shows an icescape easily identifiable on modern maps. The panoramic layout facilitates the depiction of an area with a span of about twenty-five kilometres of the horizon. The features shown in Rutter's work – ravines, glaciers, valleys, and hills – are clearly locatable on modern maps, as is the hill of Akínârssuk from where Rutter probably took his initial sketch. The accuracy of the topography is startling, and his attention to detail is evident. Although the title of the picture ('A View') and its inscribed frame, as well as its lack of bearings, suggest its maker's desire for it to be seen as a piece of art, scientific and topographical accuracy take precedence here. The picture appeals neither to the picturesque nor the sublime. Although the foreground could be said to lead the eye towards the largest glacier, a lack of novelty negates the picturesque, while the blue skies counteract any sublimity that the vastness of the ice might have suggested.

Furthermore, the picture holds relevance for modern glaciology and climate change. Taking measurements from modern maps, we can see that the glaciers have retreated significantly since Rutter's eyewitness account, with the largest glacier, Harald Moltke Brae (Sermersuak), having retreated roughly eight kilometres from its face since 1850. This clearly shows the potential of appropriate nineteenth-century visual records to be used in the accumulation of historical data to extend glaciological records further back into the past. The precision of the painting indicates that naval personnel were expected to take topographically accurate drawings and did so, to the best of their abilities. The blue skies and benign weather conditions evident in the picture would have enabled it to be executed with a high level of detail.

While a clerk like Rutter would not have been expected to make drawings as part of his job, he did receive lavish praise in Saunders's report, for his 'zealous and active part' in the expedition and was strongly recommended for promotion.[44] Higher-ranked officers, such as those leading sledge journeys in the spring of 1851 as part of the Austin voyage, were explicitly instructed to keep visual records. Spring and its lengthening days brought preparations for sledge-travelling to search for signs of Franklin and to record unexplored coastlines. Officers were expected to make measured topographical drawings on these journeys and did so with varying degrees of success. Temperatures in spring in the Arctic could, of course, be harsh, but it was occasionally above freezing. Cold affects motor skills such as finger and manual dexterity negatively,[45] and combining the rigours of sledge travel with the delicate act of drawing cannot have been easy.[46] Away from the ship, the transportation and care of fragile paper drawings must have been difficult.

In the spring of 1851, Horatio Austin gave clear instructions for his officers to draw while on these journeys: 'You are expected to make all sketches of headlands &c. that you can; and are referred to the "Admiralty Manual of Scientific Inquiry", in order that you may make such observations and collections as may prove useful.'[47] On this occasion, however, Lieutenant William Henry Browne was given more specific instructions, prior to his sledge journey along the unexplored east coast of Prince of Wales Island, which reminded the lieutenant of his particular responsibility:

> Although observations and drawings are wholly secondary to the great work you are about to aid in carrying out, yet I feel it my duty to remind you here, that the approval of your appointment to the 'Resolute' was given in full anticipation that, if opportunity presented itself, the Government and science at large would receive from your acquirements and intelligence a considerable addition to what is already known of these regions.[48]

The singling out of Browne adds an extra burden to his duties, as Austin seems to assign him the role of expedition artist. Austin stresses the importance of adding to knowledge through drawing. The 'opportunity' he refers to is that Browne is about to travel along an unexplored stretch of coastline. In his instructions, Austin makes his assertion that Browne should justify his appointment to the *Resolute* by demonstrating his known talent as an artist. The expectation that he would produce extensive high-quality visual records may have underpinned the Admiralty's hopes.

In fact, Browne's sledge report included no drawings (apart from a map), unlike those of his contemporaries, George Frederick McDougall, Walter William May, and Robert Dawes Aldrich, who may have wanted to prove their artistic talent and perhaps achieve some of the fame enjoyed by Browne after the previous expedition. Browne, when he handed in his sledge report, included this note explaining his lack of drawings:

> P.S.-I have much regret that the sameness in the appearance of the land has not afforded me that opportunity of using my pencil which I so earnestly desired and indeed expected, for, with the exception of a view of 'Cape

> Walker' and No. 1 Cape, I have no illustrations worth notice. These, should they prove fit for finished drawings, will be forwarded when completed, and the angles, &c. when copied.[49]

As Browne discovered, the topography of the Canadian Arctic archipelago lowers in elevation the further one travels west from Baffin Bay to the Beaufort Sea, and the Austin expedition, which was further westward, was not exposed to the same dramatic scenery that he had encountered during his first Arctic winter. That the landscape, to his eye, did not provide enough variation to be worth recording as he had 'so earnestly desired' and 'expected' strongly implies an awareness that – following the success of lithographs and a panorama based on drawings from his previous Arctic expedition – his visual records should now appeal to a wider general audience. As Martin Rudwick points out, this was an audience with a heightened awareness of visual culture and an interest in '"romantic" subjects such as mountain scenery'.[50] Browne's excuse also reminds us that these officer-artists were not professional artists with any degree of autonomy but were working under Admiralty orders. Furthermore, the note confirms that drawings could be begun in the field and finished aboard ship, once enough information in the form of a visual sketch with navigational information had been taken.

Browne's surviving work from the Austin expedition (1850–1) shows that he had a good knowledge of geology, which he was able to blend with his artistic talent without sacrificing topographical detail. It was this artistic skill, as well as a real feeling for the landscape, that produced art that could be easily translated into commodities, from lithographs for a scientific audience to a panorama for the average urban dweller. *Cape Walker Encampment* (Figure 1.2) displays his convincing rendering of the sandstone cliffs on Somerset Island with the whole landscape handled with a certain softness that displays subjectivity within science.[51] It is more than a topographical record; it is a response to the landscape. His accurate description in his sledge journal shows his geological knowledge: 'The cape [Cape Walker] itself is a high cliff, or rather a series of cliffs, running in a N.W. and S.E. direction, but the highest part or cape, I think, faces N.E. The cliffs are composed of red sandstone, with masses of conglomerate, and veins of dark-looking substance, like bad coal.'[52]

Although the sea cliffs, large and imposing yet almost decorative with their cascading snow, suggest both the sublime *and* the picturesque, Browne is still fulfilling a scientific role. An interest in the subjective response to the environment did not preclude scientific accuracy; indeed,

Figure 1.2 William Henry Browne, *Cape Walker Encampment*, 1851. Watercolour, 23 × 42.5 cm. Scott Polar Research Institute, University of Cambridge, with permission.

the coexistence of science and aesthetics in both scientific and literary tracts was common until the late nineteenth century.[53] The twentieth-century geologist and artist Maurice Haycock (who painted in the same area as Browne) succinctly explains this coalescence of science and art to Barry Lopez in *Arctic Dreams*: 'He talked about why one went out to paint like that. It was a conversation with the land, he said.'[54] Browne's ostensibly objective and scientific picture betrays a subjective and romantic presence, a version of the sublime that seems unthreatening, with the soft glow of the sun emanating from the horizon. This sublimity is unlike the Burkean sublime represented in many of the published representations during the latter years of the search, and seems to reach back to that earlier, more contemplative sublime. Furthermore, the presence of tents provides a scale but also points to the sledge party's place in the environment. These are not tourists in search of landscape views, excluded from their surroundings; these are the tents of expedition members, working, living, and sleeping in the environment.

If pictures were drawn in order to add to knowledge, they could also be undertaken for professional reasons that might not necessarily be strictly scientific in nature. The *North Star*, intended to supply the Ross expedition in 1849, was given explicit instructions by the Admiralty not to winter in the Arctic that year. However, the ship became trapped in the ice off the north-west coast of Greenland and was forced to disobey orders. An annotated drawing exists that may be the captain's visual evidence needed to vindicate himself. Aylen's *Perilous Position of HMS North Star*, which is clearly signed and dated, appears at first glance to be a striking example of an on-the-spot sketch.[55] The large drawing, 30.8 × 59 cm, was folded into a rough square shape 15.5 cm in width. One might be tempted to think that this was for ease of transport back to the ship from the height where Aylen recorded the scene. The grubby, stained nature of the paper, as well as the deep creases and faint pencil marks, also suggest its origins in the Arctic itself, bringing with it the elements of nature that surrounded Aylen as he drew.

However, when this piece is cross-referenced with textual evidence in the form of Saunders's report on the voyage, it is hard to see how Aylen could have risked disembarking and travelling so far from the ship to obtain the view under such conditions:

> On the 21st of September it blew a strong gale from the eastward, and the ship drifted with the [ice] pack rapidly to the westward, and at about 5 p. m. the ice was opened, in consequence of its coming in contact with an immense berg of some miles in circumference.[56]

While the elements of the picture, such as the 'berg aground', mirror those of the text, it is possible that it was in fact drawn shortly after the event as 'proof', or during the event, from a conjectural viewpoint. A corresponding sketch in the British Library entitled *Fig. 1 at 6 p. m. Position of HMS North Star, 21st Sept. 1849. Drifting in the Ice towards Berg B in a gale of Wind from SSE* depicts a plan-view, or map, of the scene.[57] Like the other sketch, the heights of the bergs, 200 feet and 170 feet, are inscribed on the paper. It seems likely that both representations were intended to consolidate Saunders's report by verifying the incident visually. After all, he had been explicitly instructed to 'carefully avoid risking all hazard of the "North Star" being detained a winter in that region'.[58] The written information, including arrows added in red ink marked 'about 3' per hour' indicating the progress of the drifting berg towards the ship, makes clear the situation. On the right-hand side of the picture, the immense mass is confirmed to be a 'Berg aground upward of 200 ft High', while the berg drifting towards the ship 'Berg Drifting about 3' per hour' is 175 ft. The use of the word 'Perilous' in the picture's caption emphasises to the Admiralty the difficulty endured by the ship and the hopelessness of her situation.

The drawing, with its inscription, tells us much about Aylen's understanding of the landscape in which they found themselves drifting. The significance and worry of the imminent danger present the event as one worthy of visual recording. Here, the aesthetics, and indeed the topography, are unimportant; it is the position of the ship in relation to the sea and ice that matters. The picture functions as a record of the condition of the ice and of the ship confined within that environment.

Personal Records and Expedition Mementos

While it was possible for subjectivity to overlap with technical accuracy at times, there were other visual records created purely for personal use; these too could hold scientific value, and the two types of drawing were not mutually exclusive. The observational skills necessary for, and tools required by, the collection of knowledge also enabled a more subjective and emotional response to the Arctic. As Erika Behrisch argues, 'the detailed demands of scientific enquiry encouraged a poetic imagination and form of expression'.[59] Focusing on three unpublished poems by three different officers, she shows how the act of exploration 'encouraged an intimacy' with the Arctic while the Admiralty's instructions 'promoted a curiosity that was equally artistic as it was scientific'.[60] As Behrisch notes, the scientific and objective recording of observations 'depended, ironically, upon an engaged sense of place and an active relation with the Arctic landscape'.[61] What Behrisch observes in the poetry of expedition members can also be seen in their visual records. The act of drawing the landscape by the observer, situated within the natural surroundings, established an intimacy with the environment. During this period, when photography was still cumbersome and expensive, people 'took' sketches in a manner not unlike how we now 'take' a photograph. This act is not necessarily always a 'taking' or appropriation of the landscape. Although Richard Sha points out that the common use of the phrase 'to take a sketch' in the late-eighteenth and early nineteenth centuries associates it with 'appropriative force',[62] Majid Yar has noted that the gaze is not only connected to power, and other possible aspects include emotional and communicative experiences.[63] Indeed, those on the Franklin searches were more likely to use the word 'make' than 'take' when referring to sketches.[64]

Drawing, along with reading and keeping a personal diary, was one of the middle-class values central to work and pleasure that transferred to shipboard life and encouraged social cohesion on expeditions.[65] While naval officers were expected to produce basic topographical drawings as part of their duties, individuals aboard ship, including non-officers, often created drawings for other reasons, such as to include in a letter, or to make sense of their environment personally and illustrate their journal, to create souvenirs for others or mementos for themselves of people, places, and events for sentimental reasons. In one instance a small vertical painting (12.3 × 3.3 cm) of the ship in winter quarters was created as a bookmark.[66]

Many of these drawings can be seen as travel sketches, made of 'places, buildings, objects, people or events seen during a journey; often made to record a particular feature for future reference'.[67] Such sketches are sometimes rough drawings, or unfinished paintings, hastily made. For example, the colour sketch (Figure 1.3) done by the surgeon of the *Resolute*, William T. Domville, on the Belcher expedition shows an unfinished sketch of a spontaneous social event.[68] The dance that took place on the *Resolute* on 8 June 1852, while the ship was anchored at Lievely (Qeqertarsuaq) on the west coast of Greenland, was described by several expedition members in writing, but Domville's sketch appears to be the only surviving visual record of the occasion.[69]

The picture shows two Greenlandic women dancing a hornpipe with British seamen while a musician, seen to the left, plays the fiddle.[70] The musician, who also played a polka and waltzes according to written

Figure 1.3 William T. Domville, *Esquimaux Dance aboard Resolute*, 1852. Pencil and watercolour, 12.2 × 17.5 cm. © National Maritime Museum, Greenwich, London.

sources,[71] is partially represented in an incomplete and unfinished sketch. Despite this, close examination reveals that he must have been a Greenlander and not an expedition member, as evidenced by his jacket and trousers. The peaked hood that he wears can be seen in another depiction of Inuit from the same place and time in McDougall's narrative of that voyage.[72] The nationality of the fiddler was not mentioned in any of the written sources and shows us how travel sketches offer additional information. The two women were named as 'Sophy and Marie' by McDougall, who praised them for their 'pretty features'.[73] The picture shows a scene at odds with the idea of an Arctic wilderness; instead it appears sociable and domesticated, aspects of shipboard life that will be discussed in Chapter 2. Furthermore, the contact between expedition members and local Greenlanders was prolonged; the dance went on until midnight according to McDougall. Emile de Bray, a French officer on the *Resolute*, records that he joined officers from the *Assistance* 'dancing with the beauties of Lievely' at the houses when he descended a mountain late that night.[74] Another dance was held the following evening, at the house of 'the fair Sophy'.[75]

Other members of the expedition were also sketching during this time. James Blair Grove, mate on the *Assistance* during the Belcher expedition, included a 'hurried sketch' of one of the women after this dance on the *Resolute* in Lievely, Greenland, to send with his letter to John Barrow.[76] An extract from this letter and an engraving of the sketch were printed in the *Illustrated London News* in December of that year.[77] At the same time, Robert McCormick, surgeon on the *North Star*, also travelling as part of the Belcher expedition, wrote to John Barrow, while the ships were at Lievely: 'I have been busy all day – and have just sketched the Mountain and Danish Settlement for you, according to promise.'[78]

There is also evidence that expedition members used their drawings and paintings of the Arctic to adorn the interior of the ship or other quarters. Some pictures, worn and battered in the archive, show pinholes at the corners, and references to the crew displaying paintings in their mess exist. Johann Miertsching, interpreter on the McClure expedition, related a visit to the seamen's mess on the *Investigator* during New Year's Day 1851 in his journal: 'After the muster we went, led by the captain, to the men's quarters, and were astonished to find a hall elegantly decorated with flags, pictures, sea-paintings.'[79] This suggests that the men had painted the pictures themselves, although as the *Investigator* was abandoned, and the men ordered to take nothing, these paintings are unlikely to survive in the archive. While a small journal could be concealed about one's person, a larger painting would have been difficult to transport.[80] Assistant surgeon Edward Adams, who served on two consecutive Arctic search expeditions, probably hung a small watercolour from his first expedition (led by James Clark Ross from 1848 to 1849) in his room at the Russian trading post of Michaelovski (Taciq), Alaska, during the winter of 1850 to 1851, evincing a desire both to have a memento of that expedition on view and to create a home from his quarters.[81]

The phenomenon of drawing and painting beyond their official duties is very noticeable on the Austin expedition of 1850 to 1851 in particular, where surviving sketches show that many expedition members were drawing for purposes other than scientific ones. The sensitive and atmospheric watercolours attributed to Robert Dawes Aldrich were not necessary for the advancement of science, nor do they add to geographical knowledge. Despite Aldrich's role as lieutenant, they do not show a definite location or coastal elevation or give any indication of latitude and longitude. Rather they seem to have been made from an impulse to give visual expression to their voyage. Two small watercolours (13 × 36 cm) show ships, presumably those of the Austin expedition, in full sail on open water.[82] The paintings do not appear to be signed or dated in any visible way. The untitled paintings have a panoramic format that lends itself well to the wide horizons visible from the deck of a ship. The watercolours have been created as a pair, one complementing the other, possibly depicting dusk and dawn. The locations represented in the paintings are unknown, but the artist appears to be concerned with using atmospheric light effects to convey a mood rather than with geographical location. They are both mounted on stiff card, and lines from a song, written on card, are glued to the mount.

Both paintings depict summer; the relatively calm weather and apparently ice-free sea suggests a location at the southern end of Baffin Bay, an area known for its pleasant weather at that time of year. It is significant, too, that no ice is represented in these private pictures, for seas choked with icebergs were an essential element of the public depiction of the Arctic. One painting (PAH0072), darker than the other, is accompanied by lines from a song suggesting the time of day. The work may depict midnight in high latitudes near the Arctic (for example, the south of Greenland in early summer). The lines of the song read: 'Or sailing on the midnight deep / Whilst weary messmates soundly sleep / The careful watch patrols the deck / To guard the "Ship" from foes or wreck.' These lines are taken from a song, called 'All's Well', by Thomas Dibdin (1771–1841), a duet that appears in several nineteenth-century books of songs.[83] The choice of lines from a song to accompany the painting suggests it was a song that was sung on board and shows the interconnectivity of cultural life aboard the search expeditions.[84]

The second painting (PAH0071) appears to depict dawn at a similar latitude and uses the melancholy lyric of the first verse of 'Isle of Beauty', written in the mid-nineteenth century by the popular poet and playwright Thomas Haynes Bayly (1797–1839).[85] The lines are: 'Shades of evening close not o'er us / Leave our lonely, bargue awhile / Morn alas will not restore us / Yonder dear & distant "Isle".' While the original lyric, part of a set called *Songs to Rosa*, seems to speak to a lover and lament their growing old and eventual death, here it seems as though the 'Isle of Beauty' is used to express a romantic nostalgia for home. Only the first four lines of the sixteen-line lyric are used here. The lines contain a misspelling 'bargue', instead of 'bark' or 'barque' in the published versions, suggesting that the lines were known by heart rather than copied from a book. Indeed, this song is listed in *Reminiscences of the Baltic Fleet* (1855) as a 'Gunroom Stock Song' in the appendix.[86] Some men and officers who served on the Franklin search did go on to participate in the Baltic Fleet fighting against the Russians during the Crimean War (1853–6).

Both watercolours are delicate and sensitive, revealing a confident artist's hand and an interest in the treatment of light. The paintings suggest the artist's awareness of the thriving genre of marine painting in Britain, particularly in a romantic sense, where the attention given to nature (the sky and sea) is as important as the ships. It is significant that Aldrich does not portray smoke from the *Pioneer* and the *Intrepid* (two steam tenders being used in the Arctic for the first time), preferring to concentrate on the more romantic idea of sailing rather than industrial progress. So too, the accompanying lyrics, if chosen by the artist, betray his romantic sensibilities. The lines suggest a need to express a textual response to the environment, turning to well-known songs to do so. Created as personal souvenirs rather than for scientific interest, the works speak of an expectation of nostalgia for the Arctic voyage in the future.

A third painting by Aldrich (Figure 1.4) depicts a paraselene[87] in the background, providing an unusual backdrop for an extraordinary sculpture exhibition.[88] The untitled scene, again undated and unsigned, is mounted on card so the reverse side cannot be seen. The snow sculpture of Britannia takes centre stage, while another sculpture of two people is visible to the right, and in the background the four ships of the Austin expedition are apparent against the imposing bulk of Griffith Island.[89] Relaxed figures on

Figure 1.4 Robert Dawes Aldrich, *Paraselene*, 1850-1. Watercolour, 17.7 × 25.5 cm. © National Maritime Museum, Greenwich, London.

the ice are scattered among the sculptures, reminding us of the walks and games on the ice that were common when the weather was fine.

Paraselenae were sometimes seen during winter and are represented in first-hand records more often than the better-known aurora borealis, which was elusive in this area of the Arctic in the nineteenth century. The scientific interest of the occurrence, however, is tempered, as the eye is drawn to the sculptures and activity on the ice. Here, the paraselene is being depicted for its beauty, perhaps in addition to its scientific interest. The atmospheric phenomenon creates another kind of impression, a mood of a still and peaceful winter environment.

Although the above pictures are the work of a lieutenant familiar with the requirements of naval draughtsmanship, lower ranks aboard ship also sketched for pleasure, as indicated by Miertsching's comment on the pictures decorating the crew's mess. William T. Mumford, carpenter's mate on the Belcher expedition, drew a series of sketches that were placed between the leaves of his diary. His small pencil sketch of the sight of a new land, *The First View of the Coast Greenland* (Figure 1.5),[90] makes visual his description of the coast as neither a sublime nor a picturesque landscape

Figure 1.5 William T. Mumford, *The First View of the Coast Greenland 21.5.52*, 1852. Pencil, 6.7 × 22.7 cm. Library and Archives Canada. Photo by author.

but as a 'frieze of broken bottles on a garden wall',[91] a line with no aesthetic or scientific pretensions but that shows his visual interpretation of the mountainous region.

The sketch, delineated by the carpenter, contains none of the scientific information expected for navigation such as bearings, latitude, or longitude. Additionally, the distant view of Greenland's mountains does not require the use of perspective. This drawing also marks what must be, for Mumford, the real beginning of the adventure, and the lines of the new horizon mark a clear distinction between the country he has left and the region he is about to enter. Mumford's drawings constitute snapshots on a journey, a record that is personal. Furthermore, the worn and damaged appearance of the paper, its unusual size and shape, as well as the use of pencil (which would be easier to use on board a ship) suggests its origins and life aboard ship, particularly as Mumford had to travel back on the *Phoenix* when Belcher ordered the abandonment of his four ships after two years in the Arctic.

It is unusual to find much material by non-officers in the archives; as Richard Harding points out, the records of officers are generally better preserved.[92] This makes the unique sketch by an able seaman, William Blakey, all the more intriguing. Blakey drew a souvenir of an incident while aboard the supply ship *Phoenix* in June 1854, under Edward Augustus Inglefield (Figure 1.6).[93] The small monochrome sketch has been mounted and framed as an elaborate keepsake. Indeed, the double frame, one around the picture and a larger one encompassing picture and text, not only commemorates the incident but also Blakey's participation in it and the role of the relief expedition in the overall search.

The main text within the frame reads: 'The Relief Expedition consisting of Her Majesty's Ships "Phoenix", "Talbot" and "Diligence" passing a remarkable Ice Berg in Davis Straits, while proceeding with Provisions and Coal for relieving Sir Edward Belcher's expedition in the Arctic Seas on the 4 day of June 1854. Height 280 Feet.' The text on the left-hand side of the picture reads 'Drawn by W. Blakey. RN', while the right-hand side bears the inscription 'In Latitude 67° 50' North'. It is possible that Blakely was familiar with engravings like *The Ice Islands* in James Cook's *A Voyage towards the South Pole* (1777) and aimed to emulate such works.[94]

The Arctic environment and positive experiences that took place on expeditions were depicted as responses and as personal souvenirs, particularly for those expeditions that entered the Northwest Passage via the high coastlines of Greenland and the eastern archipelago. But, for the expeditions that wintered in the vicinity of the Bering Strait at

Figure 1.6 William Blakey, *An Incident in the Search for Sir John Franklin*, 4 June 1854. Graphite and pastel, 6.4 × 10.3 cm (excluding frames). © National Maritime Museum, Greenwich, London.

the western entrance to the passage, the topography encountered by the ships was less striking than the high mountains and towering icebergs of Baffin Bay. There was also more extensive and prolonged contact with Indigenous peoples due to the nature of settlement along this Alaskan coast, then part of Russian America. Edward Adams, assistant surgeon on Collinson's *Enterprise* from 1850 to 1855, drew and painted regularly; his visual work includes thirty-four portraits of Yup'ik, Iñupiat, Gwich'in, Deg Hit'an, Northern Copper Inuit, and Inuvialuit, as well as birds, landscapes, and maritime views. His journal text too is filled with illustrations (on at least twenty-one pages) including drawings of Yup'ik bone carving.[95] His most striking painting is that of a young woman, painted in 1851 when Adams was twenty-seven and spent a winter ashore at the Russian fort of Michaelovski in Norton Sound, near the Yup'ik village of Taciq (Figure 1.7).

Figure 1.7 Edward Adams, *Koutoküdluk – My First Love, 1851*. Watercolour, 18.5 × 13.6 cm. Scott Polar Research Institute, University of Cambridge, with permission.

The delicate painting, *Koutoküdluk – My First Love, 1851*,[96] indicates the desire to keep a personal memento of a relationship in the Arctic. The light blue oval-shaped background frames the woman in the absence of an actual gilded frame, commonly used in the 1800s for loved ones, and is intriguing for its use in this context. Significantly, although Adams did other paintings and drawings of people, this is the only surviving picture by him on which such a frame occurs. This painted frame seems to confine the woman in a private space and implies that this was a portrait the artist created expressly so that he could gaze upon it. The size and format of the painting aligns it with the miniature portraits that were popular in the nineteenth century, which were often created as private tokens of affection. As Kristina Huneault points out, titles such as this, where the subject is named 'convey a sense of specificity quite unusual within the context of colonial representation'.[97]

Even disregarding the bold inscription and the naming of the subject, the visual format of the portrait betrays the importance of the woman Koutokǘdluk to Edward Adams. While the painting has ethnographic details, particularly in its depiction of clothing and adornment, its emotional resonance was clearly of more import for its artist. This is further seen in the blemishes on the paper, which overlay the painting and show that it was handled, and therefore looked at, repeatedly, probably during the remainder of the voyage. This distinguishes it from other portraits by Adams from the same period that are completely unmarked. Although Adams kept a journal from October 1850 to July 1851, his writing reveals little of his personal life, and this woman is not mentioned by name, likely due to the fact that he knew all written records would be read by the Admiralty on their return. However, it becomes obvious when reading the journal that it must have been written after the events took place. This is apparent when he refers to events in the future as though they have already happened. It is possible that an earlier version exists and that Adams created a second version in order to edit Koutokǘdluk's presence out of the text. Neither does he make much reference to his drawing practice in the journal, although the following incident is mentioned when two Iñupiat men travelled to Taciq: 'the chief refused to allow me to sketch his face, saying it was too ugly, & recommending his friend as a more creditable example of his countrymen'.[98] Such a statement shows clear Indigenous agency – people did not always agree to having their portraits taken – as well as signalling that they took an interest in how they were portrayed to those in more southerly latitudes. No engraving or published version of the painting of Koutokǘdluk came to light during the research for this book, and by 1856 Adams had died of typhus in Sierra Leone.[99]

The Unseen Arctic Summer

It is significant that many of the drawings created on board the search ships show an Arctic summer of calm seas and blue skies, yet the overwhelming preconception of the North, then as now, is of coldness and dark, snow and ice. This is despite the fact that not all of the expeditions wintered in the Arctic; many travelled there in early summer to resupply the search ships and returned to England in September. The dominant idea of the Arctic as a desolate and frozen realm, shrouded in mist, was the one chosen more often for representation in the public realm and reinforced the paradigm of a harsh and cold North. Heidi Hansson shows that the 'cluster of associations surrounding the popular idea' of the Arctic 'includes severe cold, distance from civilisation, dangerous conditions, barrenness and exposure to natural forces'.[100] She goes on to note 'winter is the reference point even in works set in summer' for the '(generalized) North' that is 'defined through climate and season'.[101] Chauncey Loomis argues that, although eighteenth-century writers used the Arctic 'as a source of mystery, coldness and vastness', it was not until Samuel Taylor Coleridge's 'Rime of the Ancient Mariner' (1798) that ice became associated with a type of supernatural fear.[102] Mary Shelley then used the vastness of Arctic ice to create a polar frame for *Frankenstein* (1818). That same year saw the beginning of nineteenth-century British Arctic exploration into the then-mysterious regions around the pole and publication of narratives such as William Parry's *Journal of a Voyage for the Discovery of a North-West Passage* (1821), which, despite its 'subdued and accurate' engravings and understated text, encouraged an accepted romantic image of the Arctic.[103] Parry's expedition had penetrated deep into the Northwest Passage and wintered near Melville Island, far from Inuit settlements. From such accounts, many readers formed imaginings of the Arctic that were a 'compound of fact and fantasy touched with the power of the Sublime'.[104] Caspar David Friedrich's painting *Das Eismeer* (1823) showed a nightmarish scene of huge slabs of ice crushing a ship, typifying the Arctic imaginary of the nineteenth century in northern Europe, and the icy wastes again appeared in Ann Radcliffe's poem 'The Snow-Fiend', published in 1826:

> And by the mist of livid grey,
> That steals upon his onward way.
> He from the ice-peaks of the North
> In sounding majesty comes forth;
> Dark amidst the wondrous light,
> That streams o'er all the northern night.[105]

A focus on primary documentary visual records from the search expeditions, as opposed to public material such as engravings and lithographs, yields insights into what was represented and what kinds of working methods were used aboard ship and indeed on the ice floe or land. Drawing outside was, as one might expect, practised more in summer, and direct observation from nature was associated with visual testimony, holding the power of 'truth' in a way that a report or hearsay could not. 'On the spot' is a phrase that became common in illustrated travel narratives from the late eighteenth century, signalling that the landscape was 'based on observation in the field'.[106] The methodological phrase gave authority to the picture; however, the phrase is rarely used on expedition members' drawings in the context of the Franklin searches, probably because the sketches were not intended for public view and therefore did not need their authenticity verified.[107] However, the reality of on-the-spot visual records created in the Arctic might be viewed with scepticism by critics, due to perceived intense cold, darkness, stormy weather, and other factors such as frozen watercolours and the annoyance of mosquitoes in summer. The combined archive of pictures from the Franklin search expeditions suggests a different story. Summer was certainly a time when detailed sketching could be done on the spot with relative ease, with temperatures well above freezing and calm weather often experienced. In June 1848, off the Whale Fish Islands (Imerigsoq) in Greenland, Cresswell, mate on the *Investigator* as part of the James Clark Ross expedition, wrote in a letter to his parents: 'The weather is most lovely, quite hot, the thermometer up to nearly sixty and as high as 70 in the sun. We have hardly had a drop of rain since we left England.'[108] Moreover, there seemed to be ample time to read and practise drawing: 'I have taken great pains with my drawing since I have been away . . . I have read a good deal since I have been away, and have got enough books to last me for years.'[109] Adams too, on the same expedition, recorded frequent fine and calm weather in his journal, including 'perfect' days when the sea was 'mirror-like'.[110] Such written records of the summer weather are mirrored in the visual evidence provided by the archive. The preference of expedition members for drawing on the spot is reflected in the high number of drawings that are dated between May and September.

Comparatively calm summer weather and often slow progress on sailing ships facilitated drawing, whether a measured topographical view or a travel sketch, on deck. In Baffin Bay during the summer, as expeditions made their way along the west coast of Greenland en route to the Northwest Passage, the sailing vessels often travelled quite slowly due to calm weather or loose ice. Lieutenant Walter William May recorded speeds of one-and-a-half to four miles per hour as the five-ship Belcher expedition made its way along the west coast of Greenland during the summer of 1852.[111] On this expedition, travelling as ship's surgeon of the *North Star*, Robert McCormick drew hundreds of sketches in his journals and sketchbooks, including coastal views, settlements, and events.[112] These rapid pencil sketches were sometimes annotated with written information regarding colour, using text and drawing to record the landscape and indicating that he foresaw the possibility of working them into paintings later. Mumford also created pencil sketches on the voyage (see Figure 1.5) and, later, developed these drawings, mainly of the ships in the Arctic environment, into paintings.[113]

A visit to Wolstenholme Sound in the north of Greenland by the *Assistance* and the *Intrepid* from the Austin expedition in search of the *North Star* in August 1850 provides unusual evidence of the calm summer weather that can settle in the Arctic and the burst of growth that twenty-four-hour daylight instigates in high latitudes, in this case over 1,000 kilometres north of the Arctic Circle.[114] A small watercolour by an unknown hand (Figure 1.8) shows Dundas Hill (Uummannaq) from the sea, clearly viewed from a ship's deck.[115] The torn and stained paper of *North Star Mount*, with hurried marks, points to the drawing's origin in the natural environment.

The rise in temperature accounts for the open water and grassy hills, entirely snow-free, in the depiction of *North Star Mount* by the time these ships arrived.[116] The ice-free waters are unusual to see in public depictions yet clearly possible at this time of year. Although the verdant green areas around Baffin Bay were often described in the text of narratives, they were not visually presented to the public, and plates

Figure 1.8 [Unknown Artist], *North Star Mount in Wolstenholme Sound, SE. 1.8 miles: land of a brownish purplish tinge, August 15th 1850*, 1850. Pencil and watercolour, 14 × 22.4 cm. National Library of Australia.

depicting flora were reserved for scientific tracts from expedition naturalists like Berthold Seemann.[117] Such a vision certainly did not make its way into the public spectacles of panoramas like *Summer and Winter Views of the Polar Regions* (1850) (discussed in Chapter 4), which drew the public in large numbers and represented an icy and sublime Arctic that was more in keeping with the polar imaginary inherited from the early nineteenth century.

Representations of ice-free seas are also in evidence in the work of Cresswell on the James Clark Ross voyage of 1848 to 1849. Cresswell appears to have painted in monochrome on this voyage, but as colour versions of some of the paintings exist it may be that he copied the paintings after the voyage. As officers were not professional artists, solely required to make art, they were less likely to bring their own materials on board, perhaps expecting their employer, the Admiralty, to provide the necessary materials to fulfil their duties associated with topographical drawings. Once the voyage had commenced, acquiring the materials to paint or write was an issue, even for officers. Cresswell wrote regretfully to his parents in June 1848, from the Whale Fish Islands, Greenland: 'I shall not be able to do much painting as I have not got any indigo with me, the most important colour, and the only person in the ship that has is the Asst. Surveyor, but as he has but one cake which will be barely enough for him while we are out, I cannot get a rub.'[118]

The Arctic summer predominates in his visual records from this voyage, and foliage is in evidence in some of the work. *Brown's Island Aug 1848* (16.5 × 22 cm) shows the land with sparse greenery, and the calm glassy sea in the foreground.[119] The ships, tiny and insignificant, rest against a glacial backdrop, but these details take up a far smaller portion of the picture. Adams too, on the *Investigator* with Cresswell, painted as they sailed northwards, noting in his journal occasionally that mosquitoes were found ashore, but never remarking on them being a problem aboard ship and indeed noting their absence on Brown's Island.[120] Near Upernavik, he did encounter mosquitoes, but also a wide variety of butterflies, an insect wholly at odds with the popular Arctic imaginary and also mentioned by Adams as being found on the Whale Fish Islands.[121]

The Franklin search expeditions began at a time when photography was being developed, and the most well-known photographs by an expedition member were taken by Edward Augustus Inglefield, who captained the *Phoenix* in 1854.[122] Inglefield used the new collodion process invented in 1851 by Frederick Scott Archer. Although Walter May, lieutenant on the Belcher expedition, had attempted to prepare paper for calotypes (a complicated process involving the iodising of the paper followed by the application of 'gallo-nitrate') off Greenland in the summer of 1852, the three calotype negatives and one print that exist from this expedition are attributed to Domville.[123] Photography had limited possibilities; the wet collodion process used by Inglefield was time-consuming to prepare and did not lend itself to spontaneous photography.[124] Quick exposure times were not possible until 1856.[125] The photographs by Inglefield were taken on the west coast of Greenland from June to September 1854, many of them around Holsteinborg (Sisimiut) and Lievely (Qeqertarsuaq).

Snow plays a very minor role in these photographs, which are a mixture of landscape shots and portraits of both Inuit and Danes. For example, in the photograph *The Church and Parsonage, Holsteinborg* (16.2 × 21.2 cm) the village appears in the foreground, surrounded by grassy, rocky land, and against a backdrop of snowy mountains.[126] Many photographs by Inglefield show a peopled Arctic, and his attempt to formalise his photographs and compose ethnographic portraits is shown in the use of cloth behind a photograph of an Inuit woman, posed stiffly on a chair.[127] At this time, there was no clear system in use for taking anthropological photographs.[128] This peopled Arctic again disrupts the narrative of icy wastes and assumed lifelessness.

These documentary pictures (including sketches, paintings, and photographs) of the Arctic, many of which were made in summer, do not always mesh with the Arctic of the popular imagination. Indeed, these pictures upset the dominant imaginary of the nineteenth century and perhaps even the modern imaginary; viewers expect to see snow and ice in the circumpolar regions regardless of the season. Although much of the visual recording that was practised was carried out in summer, it was possible to record visually at other times, even with temperatures dropping well below freezing. While autumn life was busy with finding secure quarters and preparing for winter, the winter itself, with its focus on shipboard domestic life, created a space in which time to draw was possible. However, these winter drawings are, understandably, less likely to be landscape views; the lack of travel in the winter darkness meant that topographical drawings for the acquisition of knowledge were not expected to be produced. While humorous illustrations for the shipboard periodicals were presumably drawn below deck, there also exist winter scenes taken in the proximity of the ship. These include drawing of events such as a funeral, meteorological phenomena like the paraselene, or simply the static appearance of the ship in the ice against a backdrop of starry skies. That sketching outside was viable in winter is indicated by Walter May's entry in his journal kept on the Belcher expedition:

> Tuesday November 30th – The day was particularly fine and towards the evening a beautiful P.[araselene] was seen. There was a complete circle first round the moon with three mock moons – and also another circle of bright light outside. The light was [. . .] grand and beautiful I attempted to make a sketch of the scene which I shall bring out with the rest of my drawings.[129]

Certainly, the sub-zero temperatures would not have allowed for the use of watercolour off the ship, but it would have been feasible to create a pencil sketch outdoors and work it up to a finished drawing or painting below deck from memory that evening when, according to Sherard Osborn, 'artists painted by candle-light'.[130] Sketches made during the summer were also passed between expedition members themselves and copied during winter, as evidenced by notes on some drawings such as that on a picture by Adams, which reads 'Glacier Bay, copied from Lieut Brown [*sic*] by candle light'.[131]

Conclusion

The primary reason for the Arctic expeditions in the mid-nineteenth century was to find the Franklin expedition or discover what had happened to it. Yet a strong scientific mandate prevailed, and officers were expected to add to knowledge by producing drawings relevant for navigation, geography, and geology. Even those with other roles on the ship could be influenced by this culture, as a clerk's precise drawings show. However, it was also possible for those recording for scientific reasons to allow their work to become suffused with subjectivity while still providing an informative visual account. Drawings were also made, on occasion, as evidence of a particular situation, such as a ship's position in the ice. In this way, circumstances beyond the control of the expedition could be explained to the Admiralty once back in Britain.

Expedition members often used their representational skills to create works of a more personal nature, sometimes eluding any scientific intent. Whether made as part of a personal record of an event or as a response to some emotional connection with an aspect of the Arctic, these non-scientific works suggest an intimacy with the Arctic environment that is associated with an attachment to people and places. Many such personal documents emit an affective tone of sentimental attachment to the Arctic and members' participation on the voyages. The sensitive portrait of Koutoküdluk (Figure 1.7) by Adams is contained in an oval frame that seeks to bring her into a realm of intimacy.

Within the archive, it is apparent that the practice of creating on-the-spot drawings, whether for scientific or personal reasons, resulted in many representations of the Arctic in summer. However, pictures showing blue skies and calm waters, even greenery, were not compatible with the public's imagined geography of a white and lifeless, threatening world. These paintings, drawings, and occasionally photographs would not be the ones that would inspire the representation of the Arctic in public formats, such as narratives, panoramas, engravings, and lithographs, formats that were reserved for the more novel aspects of the northern environment, building on the imaginary of interminable snows. The examination of scientific and personal documentary art from the Franklin searches points to a visual matrix of the Arctic that is far more complex than the peaked icebergs and threatened ships that dominated the public eye.

Although drawing was still possible in winter, the focus of the visual content often turned inwards to the ship. The arrival of that season, with its attendant darkness and colder temperatures, naturally changed the type of visual material being produced, and drawing could take on a new purpose. The practice of producing periodicals on board some ships meant that illustration, in tandem with text, became an intentional and powerful source of collective amusement for those aboard. The next chapter looks at these illustrated periodicals, examining the particular role of the visual in creating a humorous Arctic.

CHAPTER 2

'Breathing Time'

On-Board Production of Illustrated Periodicals (1850–1854)

On 2 December 1852, Lieutenant Walter May wrote in his journal aboard the *Assistance*: 'The last month has passed away pretty quickly – what with theatre, papers &c ... I am busily employed at the 2nd number of the Queens [*sic*] Illuminated Magazine – and also getting up a lot of dissolving views for Xmas.'[1] As he wrote, the *Assistance* was frozen into ever-darkening winter quarters at latitude 77° North in Northumberland Sound, deep in the then-uninhabited northern part of the Canadian archipelago. The quotation is revealing, showing how this lieutenant spent his time on ambitious pictorial projects during the winter, and points to the complex visual culture of the ship. In addition to illustrating a magazine and preparing visual entertainment for Christmas, May painted the scenery for the Queen's Arctic Theatre, which produced several plays that winter. However, the only tangible evidence that now remains of all his endeavours is the richly illustrated 'Queen's Illuminated Magazine and North Cornwall Gazette', a hand-produced monthly periodical compiled for the shipboard audience, for which he created both ink drawings and watercolour paintings.[2] It is worth emphasising that the periodical included *colour* illustrations, a novelty for the readers on board that not even the *Illustrated London News* could compete with at this time.

At least eight periodicals were initiated on British naval expeditions that searched for Franklin between 1848 and 1853. Two of the five surviving periodicals from the search expeditions will be examined in detail here. These two periodicals have been selected for their rich visual material; they contain 127 illustrations between them, which vary from small black-and-white ink drawings to full-page colour illustrations. The 'Illustrated Arctic News'[3] was created on board the *Resolute* during the winter of 1850 to 1851 and the 'Queen's Illuminated Magazine and North Cornwall Gazette' was handwritten and illustrated on the *Pioneer* and the *Assistance* during the winter of 1852 to 1853. The latter exists only as a single manuscript in the British Library. Both titles reference their visual material by using the terms

'Illustrated' and 'Illuminated', signalling the importance and significance of the visual content to the shipboard community.[4] In both periodicals, the prominence of the visual material adds a significant layer of humour and intimacy to the written material.

This chapter shows that these periodicals, particularly in their illustrations, had an emphasis on humour and used the incongruity of the situation as a source of humour, mining everyday Arctic life for comic representations. This was an effective way of negating the Arctic sublime, whether intentional or not. Furthermore, the illustrations show us a focus on the personal, individual, and more domestic side of Arctic exploration, a turning inwards from the summer landscapes produced between May and September. This resulted in an expedition narrative of familiarity and cheerfulness, in contrast to much of the published or exhibited information and entertainment that was circulating in the metropolitan sphere. The general view was that the expedition members (of the Austin voyage from 1850 to 1851) would endure 'a winter of solemn gloom, and many an unforeseen peril' before they would 'look upon merry England again'.[5] The shipboard representations challenge modern-day assumptions of the nineteenth-century view of the Arctic, as indeed they would have challenged assumptions during the nineteenth century. The prevailing modern popular narrative around polar exploration of this period still focuses on the hardships of sledge travel, the suffering endured stoically, near-death experiences, and intrepid endeavour, presenting men as both tragic and heroic. Even the titles of early twenty-first-century books, such as *Arctic Hell-Ship* (2007), *The Gates of Hell* (2009), and *Fatal Passage* (2002), which deal with the Franklin searches, contain trigger words that intensify the sublime to attract potential readers.[6] Although two of the on-board periodicals appeared shortly after as published books in Britain,[7] their version of the Arctic winter was outweighed, or tempered, by the volume of media that suggested otherwise: fixed and moving panoramas, exhibitions, newspaper articles, engravings, lithographs, and some travel narratives written by search expedition members (many of these media forms are discussed in Chapters 3 to 5).

Although it has been observed that periodical production and other cultural activities were an effective way of occupying idle minds in winter, I argue that additional important factors were at play: the periodicals were conceived of as future souvenirs, representing an important connection between expedition members and their Arctic experience. Crucially, the winter period provided the time, space, and even the authorisation to engage in such cultural practices as periodical production. However, the success of the periodicals (being reliant on their readership for contributions) depended on the presence of a lively and large homosocial community isolated from outside contact (both from Britain and from the Indigenous population). The periodicals can also be seen in the context of the wider popularity of magazines, journals, and newspapers that were increasingly consumed by a reading public. The library on the *Assistance* included periodicals like Dickens's *Household Words* (a small unillustrated weekly magazine), *Bentley's Miscellany* (a monthly humorous literary magazine), and the *Family Herald* (a weekly middle-class magazine).[8]

Several scholars have done important work on the phenomenon of periodicals aboard ships in the Arctic and Antarctica, including those of the Franklin search expeditions. Book historian Elaine Hoag draws attention to the on-board periodicals through her meticulous bibliographical work on shipboard printing, and she clarifies that only one periodical was ever printed on the Franklin search expeditions; the rest were handwritten manuscripts.[9] Hoag notes that, although small presses were generally taken on board the search ships, the press on Belcher's *Assistance* was large enough to print playbills measuring 43.7 × 34 cm and may have been a second-hand Albion or Columbian half-sheet demy press.[10] David and Deirdre Stam investigate the consumption of periodicals aboard ship, suggesting that expedition members read periodicals (including ones they produced) in order to 'bend time', giving rhythm to 'undifferentiated periods of darkness' while 'locked in the dark prison of the polar night'.[11] Erika Behrisch Elce further argues that the periodicals produced on board were distinct from those brought from home; the former had a 'richer value' and provided an 'intellectual satisfaction' that distinguished them from other activities available during the winter.[12] Periodical production, as well as other cultural activities such as theatrical performances, was encouraged as it was linked to keeping discipline aboard ship, helping to keep sailors visibly busy.[13] Crucially, Behrisch Elce observes that such productions are evidence of the importance of the social life of expeditions and argues that the shipboard periodicals, with contributions by both officers and ordinary seamen, show the importance of both reading and writing on expeditions. In their creation of a 'collective narrative', they differ from official narratives published in England, giving a far more human and domestic view of life in the Arctic as distinct from the public image of heroic self-sacrifice.[14]

Hester Blum, in her insightful essay on polar periodicals, remarks how the first Arctic periodicals focused on the 'local and intimate' despite the nationalist aims of the expeditions' endeavour.[15] Noting that the periodicals were intended for the private shipboard audience (certain articles were removed before publication), she surmises that they had a different orientation from public accounts such as book-length narratives[16] and, like Behrisch Elce, observes that they offer an 'alternative account of the experiential conditions of polar exploration'.[17] More recently, Blum argues in her compelling book, *News at the Ends of the Earth*, that polar newspapers can be seen as the 'social media' of expeditions and that the nineteenth-century Arctic periodicals 'were a mechanism for sojourners in Arctic winters to comment upon and conceptually orient themselves within the unexpected vicissitudes of polar environmental spaces'.[18]

While groundbreaking work on the on-board Arctic periodicals has been carried out by literary scholars, it is evident that what has been published understandably focuses on the text of the productions – the illustrations have not been considered in any great detail. By attending to the extensive visual material of the periodicals, we are provided with an even starker 'alternative account' of the expeditions that were perceived at the time as being 'remote from the haunts of men, beset with a thousand dangers, and imprisoned in regions of eternal ice'.[19] It is in the illuminating and immediate nature of the illustrations that the life of the expeditions becomes most visible. For example, the cross-dressing that took place for theatrical performances and masquerades is well illustrated in the periodicals, providing a closer look at this common naval practice. Thus, the visual (including its interaction with the text and verbal sketches that create a visual scene in the mind of the reader) is privileged throughout this chapter.

Shipboard Periodicals on British Naval Search Expeditions

The 'Queen's Illuminated Magazine' and the 'Illustrated Arctic News' were not the first on-board periodicals produced in the context of Arctic, or indeed polar, exploration. From 1819 to 1820, officers on William Edward Parry's expedition to the Northwest Passage wrote the 'New Georgia Gazette or Winter Chronicle'.[20] Parry's was the first expedition to deliberately spend the winter period in the ice, and he arranged for diversions, including the production of a weekly newspaper.[21] The idea was subsequently taken up enthusiastically by several expeditions participating in the mid-century Franklin search, particularly by members of the four-ship Austin expedition during 1850 and 1851, when at one stage five periodical titles were in production concurrently. These on-board productions, which were generally, though not exclusively, a feature of the winter period, were mainly handwritten manuscripts produced at weekly or monthly intervals.

While the 'New Georgia Gazette' was a text-centric production, the Franklin search expeditions' output included heavily illustrated periodicals, two of which are examined in this chapter. Unlike Parry's earlier periodical, those produced by the search vessels encouraged contributions from everyone on board and were designed to be read by, or be read aloud to, the entire ship's company. For example, John Matthews, boatswain on the *Plover*, read a copy of the 'Weekly Guy' aloud each week.[22] While the image of the British sailor is not commonly associated with reading, Blum notes that, by the mid-nineteenth century, literacy among seamen was around 75 to 90 percent.[23] The introduction to *Arctic Miscellanies* (the published version of the periodical 'Aurora Borealis' on the Austin expedition) took note that the sailors' contributions revealed a 'delicacy of imagination and a power of perception that one has great difficulty in reconciling with the honest roughness of their appearance'.[24]

The periodicals on the search expeditions were but one facet of the intense cultural and intimate domestic life that also included printing playbills and other documents, elaborate theatre productions, masquerades, concerts, and dances. The number of expeditions, combined with the popularity of cultural production, prohibits an account and examination here of some of these fascinating aspects of expedition culture. These are admirably treated elsewhere, in the work of Elaine Hoag, Heather Davis-Fisch, Heidi Hansson, and others.[25] The Arctic environment was key to the cultural production aboard ship, and its portrayal in the periodicals is characterised by an inversion of unconstructive attitudes to winter and darkness, instead representing an Arctic life that was familiar, humorous, and personal. However, these attributes were sometimes lost when the experiences of the expeditions were represented through various media in Britain, and the Arctic often became unfamiliar and depersonalised, a theme that will be further explored in later chapters. It is worth noting here an observation from a sociological and psychological perspective on nineteenth-century voyages to the Arctic: 'Much of what has been written about voyages to and in the Far North emphasises the vastness, remoteness, strangeness, and dangers of the Arctic. And yet we have many positive reactions to the Arctic in the personal journals.'[26] More recently, Lawrence Palinkas and Peter Suedfeld have observed that little research has been done on the 'positive effects' of polar expeditions, compared with 'work that emphasises negative effects and consequences'.[27]

From October, when a search ship intending to winter over was firmly frozen in the ice and the nights lengthened, a metamorphosis took place on the ship, which would last until March. In November 1852, while based at winter quarters as part of the Belcher expedition in Northumberland Sound deep in the Canadian archipelago, Commander Sherard Osborn revealingly began the first page (Figure 2.1) of the new on-board manuscript periodical, the 'Queen's Illuminated Magazine and North Cornwall Gazette', with what reads like a written sigh of relief: 'Darkness and Winter have for a while proclaimed the long night of the Arctic zone, to us a breathing time from arduous and successful labours.'[28] Winter, then, could be a 'breathing time', a time and space in which expedition members had room to pause and reflect, and to create. This breathing time was also one that allowed a 'group-spirit' to thrive on certain expeditions. As social historian Carolyn Strange has highlighted with reference to studies in present-day Antarctica, an isolated group confined in a potentially dangerous environment perceives that environment as less dangerous if the group gets along and works together well, having an overall positive outlook.[29] The use of humour in the periodicals points to a genuine cheerfulness in winter quarters when having a positive outlook contributed to a successful expedition.

Ships overwintering could be immobile in the ice for up to eleven months at a time, and this stasis provided a type of home, complete with supplies and tools that could be used for cultural activities. The periodicals and theatrical performances can in some sense be seen as the arts of immobility, encouraged by their enforced winter sojourn. Under normal circumstances, a naval ship on a surveying mission, for example, would be mobile, requiring the ongoing production of new coastal views and maps and keeping its crew and officers busy with ocean-going duties. The phenomenon of the ship in winter quarters provided a temporary stability, in contrast to the typical seafaring life that lacked 'firm geographic boundaries or stable residence'.[30] For maritime Arctic expeditions, winter quarters had the potential to become 'the heart of the emotional community'.[31] Overwintering expeditions created a space, both materially in terms of the ship as a home and psychologically in terms of the quiet time that the winter darkness provided. This provided expedition members with the space, the time, and even the authorisation[32] to partake in a variety of activities beyond the remit of searching, discovery, or science. The ship, in other words, was an entity that became a space of production, which can be regarded in terms of Henri Lefebvre's 'social spaces'.[33] Thus, 'space is both produced and productive'.[34] This space included the surrounding

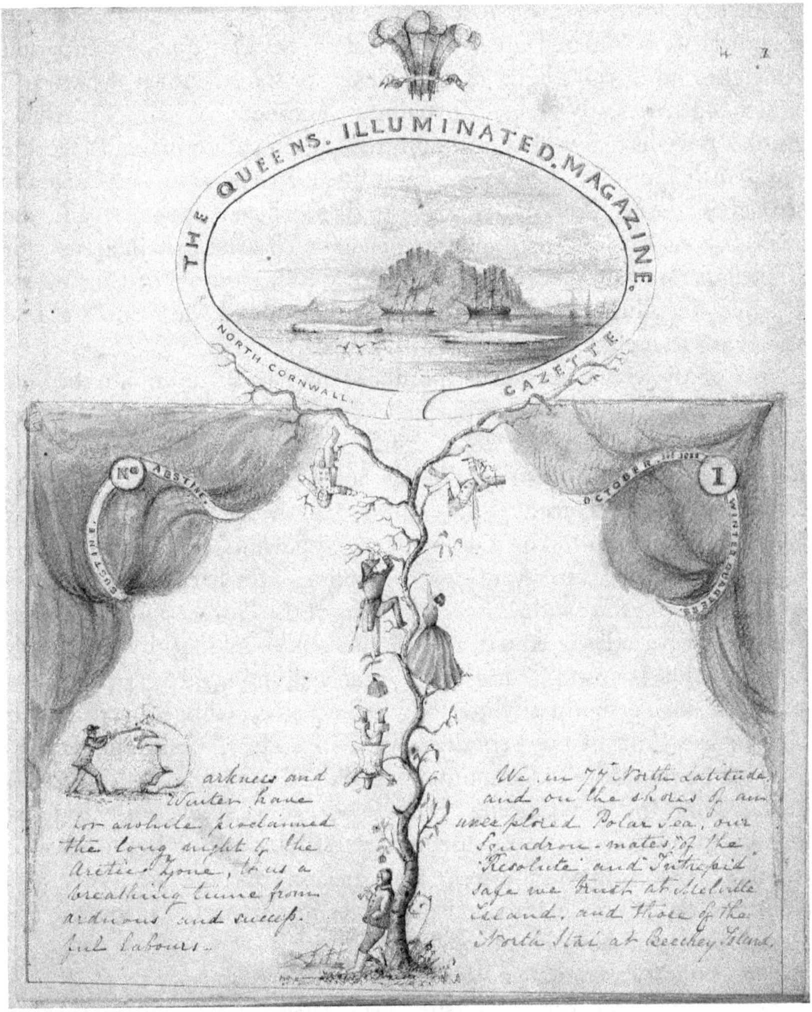

Figure 2.1 Walter William May and Sherard Osborn, title page for the first issue of the 'Queen's Illuminated Magazine and North Cornwall Gazette', November 1852. Pen and ink; watercolour, 34 × 28 cm. © British Library Board, London / Bridgeman Images.

environment and linkages to other ships in the area, making a cluster of social spaces and a cultural landscape in the Arctic, a temporary community or communities in the local environment.

The handwritten and illustrated manuscript periodicals were produced and circulated aboard, as well as between ships if a group of ships wintered together. At least seven distinct manuscript periodicals and one printed periodical were initiated on British naval expeditions during the period from 1850 to 1853: 'Flight of the Plover or North Polar Charivari'; 'Illustrated Arctic News'; 'Minavilins'; 'Gleaner'; 'Arctic Charivari'; 'Aurora Borealis'; 'Weekly Guy'; and the 'Queen's Illuminated Magazine and North Cornwall Gazette'.[35] Only the 'Weekly Guy' was actually printed in the Arctic.[36] Hoag suggests that deficient typecases and the small size of the on-board presses generally discouraged their use for printing periodicals.[37] Walter May remarked in his journal that the necessity of having illustrations in the 'Queen's Illuminated Magazine' was one reason not to use the printing press on board.[38]

Five of the eight periodicals produced by British naval searches for Franklin came from the Austin expedition of 1850 to 1851 (an astounding number of titles even for four ships). Three of those ('Minavilins', 'Gleaner', and 'Arctic Charivari') do not appear to have survived, although a satirical article copied from 'Arctic Charivari' exists in Erasmus Ommanney's papers.[39] The 'Gleaner' and 'Minavilins' (the latter initiated by Clements Markham) were two competing short-lived satirical newspapers of the *Resolute* and *Assistance*. 'Flight of the Plover' was a four-page manuscript periodical, edited aboard the *Plover* in 1848 by the ship's surgeon John Simpson,[40] and the 'Queen's Illuminated Magazine' was produced during the first winter of the Belcher expedition from 1852 to 1853. At least four of the periodicals were illustrated.[41] 'Aurora Borealis' contained a number of illustrations,[42] as did 'Minavilins'; the 'Illustrated Arctic News' and the 'Queen's Illuminated Magazine' were richly illustrated both in colour and monochrome, testifying to the importance of visuality aboard ship and the desire to create and consume visual representations of expedition members' experiences in the Arctic, including the domesticated and homely side of Arctic life.

The periodicals contained, among other things, reviews of plays and events (which, in the surviving issues, were generally enthusiastic to avoid ill feeling), poems and songs, essays on topics of interest, reports on local Arctic events such as the expedition's progress, hunting or sledging, amusing illustrations, and mock advertisements. The tone of the periodicals tends to be humorous and gently irreverent, although more serious writing (particularly related to Arctic meteorology or history) and visual material also occurs. Two of the shipboard periodicals were printed and published as books in London on the expedition's return: the 'Illustrated Arctic News' was printed as a lithographed facsimile of the same name in 1852, and selections of 'Aurora Borealis' were published as *Arctic Miscellanies: A Souvenir of the*

Late Polar Search (1852).[43] The *Illustrated Arctic News* was priced at twenty-five shillings (about two weeks' wages for a seaman at the time[44]), whereas the more modest *Arctic Miscellanies* (which had far fewer illustrations) cost twelve shillings. A second edition of the latter, which sold for ten shillings and sixpence, was published by Hurst and Blackett when it took over Colburn and Company.

The 'Illustrated Arctic News' and the 'Queen's Illuminated Magazine' are connected by the involvement of Sherard Osborn (1822–75), a key literary figure in Arctic exploration. Osborn went on his first Arctic expedition in 1850 as lieutenant commanding the *Pioneer*, a steam tender under the Austin expedition. He co-edited the 'Illustrated Arctic News' with George F. McDougall during that winter and continued his literary interests during the Belcher expedition of 1852 to 1854, when he led the *Pioneer* as commander and produced the 'Queen's Illuminated Magazine' with Walter May. He was also the manager of the 'Arctic Philharmonic Entertainments' on board the *Pioneer*.[45] He clashed with Edward Belcher to the extent that he found himself in the brig in early 1854 while the ships were still in the Arctic. After his return from his 'cruise with that rascal Belcher', he felt that his naval career was at an end: 'My naval career I feel is over. I had an object and honourable ambition once – it's gone now. I live merely for myself now and it is little I want.'[46] However, he subsequently led a fleet of ships to China in 1857 aboard the *Furious*. Samuel Gurney Cresswell, an artist and fellow Arctic officer serving under Osborn in the expedition to China, described Osborn in a letter home as 'a great man, a philosopher'.[47] Osborn wrote several books, including *Stray Leaves from an Arctic Journal* (1852), a light-hearted account of the Austin voyage that I further examine in Chapter 3.

The illustrations in the 'Illustrated Arctic News' were produced by George F. McDougall (1821–76), the second master on the *Resolute* on the Austin voyage of 1850 to 1851. He returned to the Arctic as master of the *Resolute* with Captain Henry Kellett in command during the Belcher expedition of 1852 to 1854, when he was on the management committee of the Royal Arctic Theatre.[48] Lieutenant John Bertie Cator, commanding *Intrepid* on the Austin expedition, held an unflattering view of McDougall as 'conceited, opinionated' and judged him to be 'one of the laziest fellows I ever met, a great lubber and a nervous fellow', although he conceded 'one thing he does very nicely is chart making and etching'.[49] McDougall published a narrative of his experiences on his second Arctic expedition: *The Eventful Voyage of H.M.S. Resolute to the Arctic Regions* (1857), which is also looked at in Chapter 3.

Although none of the illustrations in the 'Queen's Illuminated Magazine' is signed, all the available evidence points to Walter William May (1831–96) as the artist. May's first Arctic expedition was with Austin when he served as mate on the *Assistance* from 1850 to 1851. Like McDougall and Osborn, the Belcher expedition of 1852 to 1854 was his second Arctic voyage. He served directly under Belcher as second lieutenant on the *Assistance*, spending the first winter producing much visual material for consumption aboard the ship. His sketches that depicted the progress of the expedition were published as fourteen lithographs in 1855 after the Belcher expedition returned, and he is responsible for some of the drawings that were engraved for Francis McClintock's narrative *Voyage of the 'Fox'* (1859).[50] However, his journal kept on the Belcher voyage shows his increasing discontentment with his superiors during the second winter, complaining of the 'depraved state' of Belcher's mind.[51] According to naval historian Andrew Lambert, Belcher had a 'tempestuous, confrontational leadership' style that was not suited to the 'close confinement of a polar winter'.[52] May went on to leave the navy and became a successful marine painter and watercolourist.[53] I examine more of his paintings and drawings from the Arctic, as well as his lithographs, *A Series of Fourteen Sketches Made during the Voyage up Wellington Channel*,[54] in Chapter 5.

Both the 'Queen's Illuminated Magazine' and the *Illustrated Arctic News* are large-format folio productions measuring over 30 cm in height.[55] This is in contrast to many weekly publications, such as the *London Journal* and *Chambers's Edinburgh Journal*, but is more similar to the *Illustrated London News*. It is conceivable that the large folio format was chosen for use within the domestic space of the ship (reading aloud to a group) and so that illustrations could be shown to full advantage. The *Illustrated Arctic News* facsimile, in its title, played on the *Illustrated London News*, although this was not necessarily reflected in its contents. It contains seventy-two monochrome and colour illustrations on fifty-six pages. In the absence of access to the original manuscript version (which I was unable to locate),[56] the discussion here focuses on the lithographed facsimile. The original periodical was produced monthly, with five issues in total, beginning on 31 October 1850 and ending on 14 March 1851. Four numbers and a supplement of the 'Queen's Illuminated Magazine and North Cornwall Gazette' appeared from November 1852 to February 1853, during the first winter that the Belcher expedition spent in the Arctic. No periodical was produced during the second winter.[57] Magazines, produced weekly or monthly, proliferated during the Victorian period.[58] The term 'magazine' implies flexibility in the contents and a product more likely to entertain rather than inform its special-interest readers.[59] The first part of the title ('Queen's Illuminated Magazine') indicates variety and colour illustrations, yet the specificity of 'North Cornwall Gazette' alludes to a type of newspaper intended for a local and immediate readership, in this case the isolated audience on the *Assistance* and the *Pioneer*. The author of the printed advertisement for the magazine did not expect to have any news to deliver owing to their isolated situation, explaining the use of the term 'magazine' thus:

> The frozen state of the roads, the injury caused to the Electric Telegraph by the recent storms, and frost, and the breaking down of the Transatlantic line of steamers obliges us to depend, for foreign, and country intelligence, to the sole source left us of Balloons and Carrier Pigeons. We feel therefore, driven to adopt the term Magazine. Though we do not in doing so, aim at any very lofty position, in the catalogue of such publications.[60]

This kind of joking about their isolated situation sets the tone for the entire production, using the Arctic environment as a source of humour. Issues were between twenty and thirty-two pages in length, and the supplement was ten pages long. The January number, including its supplement, was an impressive forty-two pages in length. The magazine was heavily illustrated, containing fifty-five distinct illustrations, fourteen of them in colour, in a total of 103 pages. The 'Queen's Illuminated Magazine' had almost double the number of pages compared to the published version of the 'Illustrated Arctic News', suggesting either that it had a larger number of contributions, or that a significant amount of material was excised from the latter prior to its publication.

While the identity of the illustrator is not explicitly mentioned in the magazine and the pictorial material is unsigned, many of the drawings and paintings show May's confident hand, and some compositions are similar to his lithographs.[61] Osborn and May had been friends on the Austin expedition, and McDougall, Osborn's previous artistic collaborator on the 'Illustrated Arctic News', was then several hundred miles to the east, in winter quarters off Melville Island with Kellett's arm of the squadron. In May's journal, written aboard the *Assistance* that winter, he mentions the illustrations on several occasions, lending weight to the certainty that he is the main illustrator. Initially, he remarks that the 'printing press will not allow us on account of its size and a want of type to print the paper – besides the illustrations will have to be done which could not be put in if printed'.[62] May's comment signals the importance of the illustrations, which were conceived of as being integral to the production. On

8 November, he wrote: 'I finished the illustrations of the paper yesterday',[63] indicating that he is the illustrator of the 'Queen's Illuminated Magazine'.

In the early 1850s, illustrated periodicals, such as *Punch*, *Illustrated London News*, and *Reynold's Miscellany*, were well established yet the mainly wood-engraved illustrations were still regarded as a novelty. Black-and-white illustration became more widely used in the 1860s, but it was not until the 1880s, when printing techniques advanced, that illustration became commonplace.[64] So, the illustrated on-board periodicals, particularly the fact that some included colour illustrations, must have appeared novel to their Arctic readers, given their representation of the expedition members' own lives in the Arctic. The visual dimension is therefore of huge importance to both these periodicals. While in some instances the pictures in these periodicals serve to illustrate the text, thereby visually elaborating, in other cases the visual is prioritised and the text takes its lead from the pictorial representation, thus showing the centrality of the illustration to expedition life and the importance of pictorial self-representation of expedition members. The front page of each issue is well illustrated, and the placement of illustrations at the top of the page shows their dominance.

Larger watercolours in the 'Queen's Illuminated Magazine' (measuring up to 16 × 23.5 cm), such as *Monday Evening the 16th August 1852* (Figure 2.2), were created separately and affixed to the manuscript.[65] These appear on the title page of three of the issues and show the appreciation and recognition of May's talents, which appear to have motivated some of the written material in the magazine. For example, the author of the text below *Monday Evening* undoubtedly saw May's watercolours from the summer season, which triggered his written work about events that had taken place six months previously. *Monday Evening* shows a benign Arctic with blue skies. Its composition is balanced with a delicate use of colour and confident, loose lines. The written piece takes the picture as its starting point: 'The calm and [Arctic?] summer like scene above depicted will recall to most of our memories the beautiful evening in which Her Majesty's Ship "Assistance" and steam vessel "Pioneer" reached Cape Becher.'[66] The picture is later revisited as a reference point in the same article, highlighting its centrality to the text: 'However, to return to our frontispiece, within twenty four hours'[67] and 'Those who look upon the somewhat sunny scene the Illuminator has given'.[68]

Likewise, the 'Illustrated Arctic News' regarded visuality as a core part of the production. The full-page colour illustration (42 × 26.8 cm) by McDougall showing a masquerade ball on the *Resolute* in December 1850 takes a central position in an issue of the *Illustrated Arctic News*. It depicts sixty characters at the occasion, in full costume, making it possible for those

Figure 2.2 Walter William May, *Monday Evening the 16th August 1852*, 1852. Watercolour, 13.8 × 22.5 cm. 'Queen's Illuminated Magazine and North Cornwall Gazette'. © British Library Board, London / Bridgeman Images.

involved to recognise themselves (as recorded by the 'Aurora Borealis') and showing that the visual element in the manuscripts is integral and paramount:

> With a feeling of pleasure not altogether devoid of vanity, each saw, as in a mirror, a true representation of himself, figuring in the joyous crowd.

> We thank the artist for his able efforts, and with the hope that other opportunities may present themselves to exercise his faithful pencil we loudly cry 'Success to "'The Illustrated Arctic News".'[69]

At least eleven men are shown dressed as women in the picture. Expedition members regularly dressed as women to take female roles in the on-board theatre productions and also for masquerades or other occasions when fancy dress was approved. In the absence of actual women in the domestic space of the ship, the men also took to making elaborate costumes on survey expeditions and in connection with the Franklin searches.[70] The practice of cross-dressing, particularly female impersonation on the stage and in music halls, was popular in the nineteenth century, and men often played the parts of women in amateur theatricals.[71] As Davis-Fisch points out, in the play *Zero*, which was written and performed during the Austin expedition (1850–1), the two female characters, Daylight and North Polar Star, are 'aligned with domesticity and sentimentality, locating desire as something to be deferred until the return home'.[72] However, the figures in McDougall's picture card have very 'womanly' bodies, with obvious bosoms and narrow waists, and even dance with men who gaze at them fondly (at the bottom of the picture card). These women do not appear to fit into the category of the pantomime 'dame', who was a comedic 'ugly' woman without sexual allure. In the picture, at least, attempts to imitate idealised femininity, a tradition that became associated with homosexuality, are evident.[73] The term 'homosexual' did not enter discourse in Europe and America until the last third of the nineteenth century.[74] In the 1870s theatre was linked with transvestism and homosexuality and by the early twentieth century 'cross-dressing signified sexual-outlawry'.[75] However, the presence of so many men dressed as women in the on-board periodicals shows its wide acceptance as part of naval culture in the mid-Victorian period.

Arctic Humour

Winter in the Arctic, particularly in the uninhabited parts of the northern Canadian archipelago, was a time for turning inwards, a time when the focus naturally shifted from the wider natural environment and towards people and personalities on board the ship. The illustrated periodicals, with their heavy emphasis on humour and the human figure in everyday situations, reflect that seasonal change in a way that visual material

Figure 2.3 Walter William May, *Scene from The Rival Lovers*, 1853. Pen and ink, 10.7 × 20 cm. 'Queen's Illuminated Magazine and North Cornwall Gazette'. © British Library Board, London / Bridgeman Images.

produced for the general public did not. The illustrations concentrate on the personal, the individual, the domestic, and the social – revealing the proximity of personalities and the social space of the ship during the winter – whereas the Arctic of popular imagination in the metropole was largely expressionless, peopled only by individuals who were indistinct figures in a vast Gothic landscape.

This intimate Arctic space was not one with which the general public, used to scanning the wide Arctic horizon of panoramas and of the geographical imagination, was familiar. When figures were included in graphic material for public consumption, for example in the panorama *Summer and Winter Views of the Polar Regions* (1850) or lithographs like those of May and Cresswell, they tended to be engaged in outdoor masculine physical activity.[76] The Arctic of the periodicals was not sublime, heroic, or astonishing, but familiar, personal, and sociable. The periodical illustrations produced for consumption on board, some of them bursting with characters, emphasise the sociability and warmth that could exist on and between ships.

Close scrutiny of the hand-drawn illustrations in the 'Queen's Illuminated Magazine' reveals a myriad of smiling faces, like those in the audience watching *The Rival Lovers* at the 'Queen's Philharmonic Society' on *Pioneer* (Figure 2.3).[77] It is furthermore interesting that this view is shown from backstage, looking out at the audience, perhaps where Walter May, who painted the theatrical sets, would have stood, and demonstrating not only the centrality of the theatre to Arctic life but also the social aspects of the gathering that went on until around midnight.[78] The inclusion of the backstage figures on either side of the scene marks out the theatre performance as something that was produced as well as performed. By showing the event in this manner, he creates a drawing as much for those who were involved in the preparations as for the audience, emphasising the shared intimacy of Arctic life.

Another pen and ink illustration in the 'Queen's Illuminated Magazine', captioned *Now! o—ne, two—o, three – Haul!* (11.5 × 9.3 cm), shows two seamen attempting to help an actor close a dress for the theatre production.[79] Here, not only the idea of a crewmember in women's clothing provides amusement, but the fact that the seamen on either side, used to a type of hauling more maritime in nature, are now acting as maids for the 'lady'. This good cheer and vibrant social life that was possible during winter on well-organised expeditions (ill-prepared expeditions more often suffered from ill-health and cold and were prone to disaster) was noted by Clements Markham, midshipman on the Austin expedition from 1850 to 1851, who concluded privately in his journal that 'it became evident that all were determined to pass a regular jolly winter – in fact the dinners, parties, & c, were rather carried to an excess, & left little time for anything else'.[80] Markham did not include this comment in his published version of events, where he was far more restrained in his descriptions of amusements, emphasising that they were 'absolutely necessary ... to drive away ennui that might otherwise have seriously injured both the bodily and mental health of the Expedition'.[81]

The space and time provided by the winter allowed for the indulgence in humour with small in-text drawings created for the periodicals. Both illustrations and text in the 'Queen's Illuminated Magazine' show an overwhelming emphasis on the comic with a gently satirical edge. Articles and illustrations include mock advertisements, long letters of feigned outrage, and cartoons. These can be seen to combine to form a type of product similar to Victorian comic magazines that had become popular in the 1840s, such as *George Cruikshank's Table-Book* (1845), *Punch, or the London Charivari* (1841), and the *Man in the Moon* (1847). Two of the periodicals from the search expeditions ('Flight of the Plover or North Polar Charivari' and 'Arctic Charivari') even used their titles to reference *Punch*, which in turn took its inspiration from the French satirical magazine *Le Charivari*.[82] Indeed, *Punch*, which was hugely influential during this period, was, at the time of the departure of Franklin in 1845, 'the most talked-about and enjoyed periodical of its time'.[83] *George Cruikshank's Table-Book* (1845) was included in the on-board library of the *Assistance*, and its steel and wood engravings may have provided some of the inspiration for both text and illustrations in the Arctic periodicals.[84] The *Table-Book* included contributions by members of the *Punch* staff and featured essays, sketches, and comic tales, while satirising social issues.[85] The style and layout of the drawings in particular is comparable. The figures of women that appear in the Arctic periodicals could well have been based on those in this book. Indeed, on Rochfort Maguire's expedition, in 1852, the newspaper's editor, John Simpson, was given Cruikshank engravings by John Barrow to accompany the 'Weekly Guy', a projected periodical of the *Plover*.[86] Crucially, where the comic content of the 'Illustrated Arctic News' and the 'Queen's Illuminated Magazine' differ is that they satirise themselves (by making fun of their situation) and not other sectors of society.

In terms of the visual content of both periodicals, a substantial proportion of the illustrations is based on humorous interpretations of, and responses to, their situation. Twenty-seven, or over one-third, of the illustrations in the *Illustrated Arctic News* take a humorous angle, having expedition members as their primary subject matter. Fourteen of the drawings, less than a fifth, are landscapes or seascapes, and only one drawing shows any explicit danger. Over half, or twenty-nine, of the fifty-five illustrations in the 'Queen's Illuminated Magazine' are humorous, and the majority of them place an emphasis on domestic Arctic life focusing on figures, both human and animal, rather than on landscapes. One-fifth of the pictures in the 'Queen's Illuminated Magazine' are landscapes or seascapes. The amount of humorous drawings far exceeds any other category, such as seascapes or icescapes (ten), maritime subjects (eight), meteorology (five), landscapes (three), or dangerous situations (three).

The illustrators and writers repeatedly use personification, a recurring element of Victorian humour,[87] but in a way that is specific to the Arctic environment. For the expedition members, particularly those who had been on more than one expedition, the Arctic was not sublime, it was familiar. As Ann Colley discusses with respect to tourism in the Alps after the mid-nineteenth century, familiarity erodes the sublime experience.[88] Thus, factors like extreme cold, meteorological peculiarities, and wild animals were part of a domestic life that was easily mined for comic effect. This also had the result of upending any sublimity in these collective narratives. The illustrations in particular broke the Arctic down into manageable portions of humour, with personification being repeatedly employed; the sun ('this red-faced beggar'), cold, and the aurora borealis were all given animated roles, presenting them in 'homely terms', as Max

Keith Sutton notes was typical of Victorian comic personification.[89] Furthermore, their Arctic world, and their curious existence, while perhaps often spoken of as barren and dreary, was bountiful for the exploitation of humour.

As the anthropologist Mary Douglas shows, the joke form depends on the social situation. Furthermore, the joke within a social structure says something pertinent about that structure.[90] And what Richard Altick writes about *Punch*, the influential British comic periodical that began in 1841, can equally be applied to a consideration of humour, particularly visual humour, in the Arctic periodicals: 'Comedy is not necessarily incompatible with truth; signifiers in the comic mode, caricatures above all, may tell us as much about the signified as do more serious ones.'[91] In fact, the on-board periodicals can be seen to confirm and help build the identity of those on board.[92] The humour in the periodicals could best be enjoyed and appreciated by the on-board audience as a distinct social group. This humour, in both illustrations and text, simultaneously stressed specific concerns of the group, much in the same way that modern subcultural comics do, 'without directly attacking or offending any particular group in the subculture'.[93]

While satires in the second half of the nineteenth century made fun of tourists seeking the sublime in the Alps,[94] the writers and illustrators of these on-board periodicals satirised themselves, the idea of terror, and the metropolitan Arctic imaginary. The perception of the Arctic's association with the sublime and with horror was gently ridiculed in a lengthy piece in the 'Queen's Illuminated Magazine' when the author described a stormy Arctic night during the expedition. As 'fresh reports of . . . narrow escapes from being frozen to death or carried off by lean and hungry bears' were brought in, 'the good ship trembled to her keel', and the crew, with 'horror-stricken countenances', faced 'the inevitable doom of being frozen to death'.[95] The narrator was then momentarily awakened by the bedclothes falling off before being later roused by his steward to discover it was a nightmare: 'Reader I had had tripe for dinner the day before, & all this was a nightmare. May not this account for much we read of? Crackling Auroras, Bears' Nests.'[96]

At the top right of the essay, a small note reads 'A <u>quiz</u> upon the Voyage of the Terror 1837', referring to the expedition to discover the Northwest Passage commanded by George Back. This expedition, which had spent a winter under severe pressure from the ice, had been described in Back's *Narrative of an Expedition in H.M.S. Terror* (1838). The phrase in the magazine, 'the good ship trembled to her keel', recalls

Figure 2.4 Walter William May, *Figure in Sealskins*, 1853. Watercolour, 5 × 6 cm. 'Queen's Illuminated Magazine and North Cornwall Gazette'. © British Library Board, London / Bridgeman Images.

Back's narrative in which the *Terror* frequently trembled under pressure from the ice, including 'some severe shocks, which made her tremble to her very keel'.[97] In addition, the quartermaster tells the narrator of the essay that it was the 'most awfullest night, Sir, I ever saw, just like them they had in the xxxxxx'.[98] This lengthy contribution to the magazine, which includes a reference to an 'Ancient Mariner' who speaks in 'sepulchral tones',[99] was illustrated with a watercolour *Snow Drift Feb, 15/53* and a small drawing of a hapless-looking figure dressed in his sealskins (Figure 2.4): 'Suddenly I heard a voice from the Crow's Nest, sounding above the tumult – "They smells them there sealskins and they're a coming!"'[100]

The grim Arctic calamity was described in a visual way, creating a scene of horror that was exaggerated for comic effect. A polar bear with eyes glaring like a 'whale-oil lamp' and the 'champing of her tusks . . . like the clash of a portcullis' ran at the narrator.[101] The Arctic sublime – the perception of polar exploration in the public mind – is here used before it is abruptly upended by a simple case of indigestion. This ultimate overturning of sublimity appears to poke fun at Arctic exploits and the idea of the heroic explorer, here merely a slumbering officer whose bedclothes have fallen off. Furthermore, the writing mocks the narrative of George Back while the inclusion of the 'Ancient Mariner' alludes to Coleridge's poem and its association of ice with supernatural disaster. The scenes described are all the more interesting given that, by the time this piece was written, Franklin had been missing for seven years. Moreover, three graves of Franklin's men were discovered in the summer of 1850; indeed, the phrase 'horror-stricken countenances' brings to mind the photographs of the dead men in the book *Frozen in Time: The Fate of the Franklin Expedition* (1987).[102] It is unlikely that the essay by 'A Dreamer' would have remained had the 'Queen's Illuminated Magazine' been published, particularly as the fate of the Franklin expedition had been uncovered by the time Belcher's expedition returned to Britain in 1854.

Mock advertisements were also used to see the funny side of things. For example, in the first issue, an 'adventurer' placed a 'Wanted' advert seeking a new home near London with a 'cheerful prospect embracing at least a dozen trees, some grass, and one cow'. The advertiser was open to an exchange with 'any gentleman seeking solitudes and a bracing climate, both of which are procurable by enquiring at No. 2 Pioneer Plains'.[103] The language is deadpan; the improbability of just 'one cow' appearing in a pastoral vista underscores the ridiculous nature of the request.[104] Although the page itself is devoid of illustration, the advertisement uses nouns associated with the picturesque, such as 'prospect', 'trees', and 'cow', thus invoking the image of a rural English landscape painting. The advertisement appears to ridicule the idea of the picturesque landscape whereby elements of the real environment could be selected and placed at will to form a pleasing picture. The address given, 'No. 2 Pioneer Plains', plays on the ship, the *Pioneer*, frozen into the sea-ice, but also underlines the relative stability of their situation. Such use of invented addresses was a common feature in contributions to the 'Illustrated Arctic News' and the 'Queen's Illuminated Magazine'. It established a locality in the Arctic ice, combining personalities, landscape features, and ships. An advertisement that appeared in the final issue marked the coming season of arduous travel, jarring against the image of the heroically sublime endeavour of hauling sledges to search for Franklin: 'Wanted by a person about to proceed on a long journey – a good pair of legs – his own being somewhat out of condition, and that great distinguishing feature, the calves, are supposed to have left for Cape Farewell.'[105]

The small ink drawing *The Defeat of Zero!* (Figure 2.5) shows the Arctic winter personified in the hooded and cloaked figure of Zero, who carries an oversized barometer with icy fingers, much like Death carrying a scythe,

Figure 2.5 Walter William May, *The Defeat of Zero!*, 1852. Pen and ink, 4.9 × 11.5 cm. 'Queen's Illuminated Magazine and North Cornwall Gazette'. © British Library Board, London / Bridgeman Images.

being pursued by four enthusiastic casks of beer. They jostle each other, running after Zero on improbably spindly legs as the foremost cask hits the grotesque figure with a tankard. The incongruous nature of the rotund barrels racing after a cloaked figure depletes the possibilities of the sublime. Here, cold and possible death are defeated by animated everyday objects. The cartoon was probably inspired by the on-board brewing endeavour that was set up,[106] while the figure of Zero was developed from May's experience on the Austin expedition, when an entire pantomime, *Zero*, was written and performed especially for the character.[107]

The 'Illustrated Arctic News' too made use of humour to amuse its readers. A comic sketch shows an expedition member regarding the snow sculpture of Britannia on the ice and is captioned *Oh! So You Rule the Waves – Do You. Ah! A devilish easy berth you'll have of it this winter.*[108] The illustration and its caption show the ability of the naval officers to make fun of imperial aspirations, the British Empire, and the nature of the Royal Navy. In fact, it seems to undermine the authority of the Royal Navy by immobilising it. It also shows the elaborate sculptures of a 'Walhalla',[109] a humorous allusion that refers to the hall of fame containing busts and plaques of significant German-speaking people, which was completed in Bavaria 1842 by King Ludwig I and had been admired in England.[110] In the background, the ships covered in for winter, connected by a series of pillars, show the familiar Arctic scene of winter quarters. Scenes of ships in their winter quarters were commonly associated with the romantic sublime, yet here any possibility of the sublime is overturned by the joke.

The reality of day-to-day life in the Arctic may have inspired the cartoon *Northern Sports – Tumbling on the Ice* by McDougall that appears in the March issue of the *Illustrated Arctic News*. On 6 February 1851, William Henry Browne slipped on the ice and fractured his radius. The surgeon, Abraham Bradford, kept detailed notes on the lieutenant's progress and administered opium to ease any pain.[111] Once again, the illustration uses circumstances and local events from which to draw a comic response. The men, rather than conquering the Arctic environment, are transformed into helpless children who flail about in a bizarre situation. This emphasis on humour for collective, yet selective, viewing (those participating in the expedition) is far removed from the romantic introspection of some of the watercolours and drawings, like those of Robert Dawes Aldrich discussed in Chapter 1, which may indicate the more personal use intended for the latter. The emphasis on comedy in the periodicals suggests that they were an outlet for comic interpretations of their situation, something that, given the gravity of the search for Franklin, neither scientific articles nor published narratives allowed. The writing and drawings in the periodicals represent another form of personal response to the Arctic, something for which there was no room in official reports.

Although the idea of the periodical was brought from Britain, the Arctic expedition's periodical had its own separate identity, and the ice-bound ship in the Arctic environment proved to be a rich source of humour that was prioritised in both text and illustration. The incongruous nature of the situation in which the men found themselves provided a productive seam of inspiration for a form of humour that effectively negated the sublime. Indeed, the terror of the sublime was even gently mocked as a side effect of indigestion in one article. The British naval Arctic expedition members viewed the Arctic as a familiar place, one where novel aspects of the unfamiliar environment became everyday normalities to be incorporated and used for humorous intent in the periodicals.

Function and Readership

Both Stam and Stam and Behrisch Elce discuss the reading of periodicals as tools of naval discipline, intended to keep potentially idle hands occupied over the winter period in the Arctic.[112] As Behrisch Elce notes, this was only one function of their production and the act of writing the periodicals was as important as reading them.[113] Although captains may have encouraged activities such as periodical production and theatre performances to allay boredom, these practices were by no means universal on the British naval expeditions during the Franklin search. Captain Robert McClure's ship, the *Investigator*, spent three winters in an area that can be described as isolated, even by Arctic standards, on the north coast of Banks Island from 1850 to 1853. McClure used lashing to enforce discipline, and, although a seamen's school was organised during the first winter, little else was done to alleviate the boredom. During the third winter, with provisions running short, Johann Miertsching (the expedition's translator) writes that 'everyone has learnt knitting or crocheting to shorten the long fearful hours'.[114] There are no records of members of the Ross expedition from 1848 to 1849 engaging in performance, and neither was a periodical produced, although Edward Adams, William Browne, and Cresswell painted during the voyage.[115]

Additionally, the reality of creative practice, and periodical production in particular, seemed to depend on the presence of certain driving actors. For example, on the Austin expedition, which had five periodicals, there were key figures, such as Osborn, McDougall, Ommanney, Markham, Frederick Krabbé, Charles Ede, and William Dean, who were instrumental in initiating and sustaining cultural activities. Expedition members on the four ships of the Austin expedition were able to feed off each other's enthusiasm and verve, exchange ideas, and cooperate in projects while also lending a competitive element to activities. The periodical 'Minavilins' was established on board the *Assistance*, so as not to be outdone by 'Gleaner' on the *Resolute*. 'Minavilins' intended to 'keep a sharp watch on the *Gleaner*', whose editors remained incognito.[116] It seems to be the case that a creative cluster, consisting of several talented and enthusiastic individuals, formed on this expedition, and that this grouping drew in other expedition members, lending creativity a contagious effect and leading to the extensive cultural production. Unlike metropolitan periodicals, the Arctic productions generated and benefited from instantaneous reader reaction within the close-knit community aboard ship, which must have provided additional stimulus for contributors and editors.[117]

Neither was periodical production confined solely to winter, as has been generally assumed. While the production of the periodicals was, like theatre, primarily a winter pursuit, they were not always limited to that season from November to February.[118] The four issues of 'Flight of the Plover' that survive are dated from 1 March to 1 June 1848.[119] The first issue of 'Aurora Borealis', the newspaper of the *Assistance* on the Austin expedition, was produced in June 1850, followed by monthly issues from

September 1850 through to March 1851, with a final issue produced in September 1851, showing us that interest in creating and reading the periodical could go beyond the mere necessity of amusing and occupying the crew members during the three darkest months of winter from November to February.

Given the tone of the 'Illustrated Arctic News' and the 'Queen's Illuminated Magazine', one can only speculate on the probably more extreme contents of two short-lived satirical newspapers of the Austin expedition, 'Minavilins' and 'Gleaner', which were quashed by superiors.[120] The 'Gleaner', edited anonymously, had 'a humorous tendency', and its second issue included 'a scurrilous . . . attack on one of the officers of the expedition'. The answering article in 'Minavilins' used 'scathing satire' and 'several humorous illustrations' in its reply.[121] Hence, the quashed periodicals likely crossed the fine line between humour and offensiveness. Perhaps originally conceived as an occupation for potentially idle hands, the periodicals could in fact become vehicles for veiled criticism and even open hostility.

The Austin expedition of 1850, with its exclusively homosocial environment in winter quarters far removed from any contact with Indigenous settlements, is remarkable simply for the quantity of surviving documents in the archive that pay testament to the enthusiasm of its members for engaging in periodical production and other cultural activities. As Blum notes, in the nineteenth and early twentieth centuries the percentage of on-board periodicals produced on polar expeditions was far higher than on non-polar voyages.[122] Even accounting for the material that is unlocated (at least three shipboard periodicals), a large amount of material remains. One distinguishing feature of this expedition in the winter of 1850 to 1851 was its location near Griffith Island, in the centre of the Arctic archipelago and not in the vicinity of any Inuit settlements.[123] As well as the four ships comprising the Austin expedition, there was a group of other ships wintering comparatively close by, in nearby Assistance Bay, some twenty miles away or a four-hour journey by dog-sledge.[124] These were the *Felix* with its unmanned tender, the *Mary*, captained by John Ross, and the *Lady Franklin* and the *Sophia*, led by William Penny. This relative proximity led to a cross-fertilisation of ideas between ships and expeditions. For example, *Arctic Miscellanies* contains letters from Assistance Bay, including an account of their own Royal Cornwallis Theatre, while a seaman from Penny's expedition provided interval entertainment during the final performance at the theatre on the *Assistance*.[125] The ships of the Austin expedition were frozen in for eleven months, and it is possible that this exclusive homosocial world, without the benefit of social interaction with an Indigenous population, was a significant factor in its intense cultural production.

With the right conditions, especially with the right captain, the Arctic winter became a refuge in which expedition members could experiment and express themselves. The social aspect of so many personalities gathered together could be inspiring: 'If a James, Ainsworth or Trollope could only spend a winter with us, what [Honey] they might store up for four volumes fictions; indeed so rich a harvest, that we almost feel tempted to begin something with an Esquimaux Title or "How I found myself Frozen in"!' exclaimed the supplement of the 'Queen's Illuminated Magazine'.[126] This allusion to the contemporary prolific and popular authors who wrote three-volume novels, George Payne Rainsford James,[127] William Harrison Ainsworth, and Frances Milton Trollope,[128] suggests that a rich and intricate web of personalities and activities was part of life during the winter.

Conversely, ships that wintered near Indigenous communities could have plenty of informal social interaction and correspondingly less time for activities such as writing for a periodical. Expeditions such as that of the *Plover* (1848–54) benefited from Iñupiat who hunted, shared their knowledge, and created winter clothing. Rochfort Maguire on the *Plover* relates how exhausted he felt at the end of a long day of social interaction with Iñupiat aboard ship.[129] Although the *Plover* did, during its six-year stay in the Arctic, produce two short-lived periodicals, these were both very small productions that lacked illustrations. The 'Weekly Guy' was a small (20 × 16 cm) four-page production that only lasted a few months due to a lack of contributions. The *Plover*, under the command of Thomas Moore from 1848 to 1852 and captained by Maguire from 1852 to 1854, spent its first winter near the people of the Chukchi Peninsula on the west side of the Bering Strait and subsequent winters near Iñupiat communities of what was then Russian America. Lieutenant William Hulme Hooper's narrative, *Ten Months among the Tents of the Tuski*, details extensive contact between the ship and the local population from 1848 to 1849, while Maguire's journals show that Iñupiat were on board, apparently at their own request, almost every day during the winters of 1852 to 1854.[130] Consequently, it seems that social interaction with their neighbours took up much of the crew's attention and spare time. This included consorting with the local women, as several sources indicate.[131]

Hoag has observed that, while fewer plays were performed on the *Plover* than on some other Arctic ships, more masquerade balls and dances took place.[132] For example, in February 1853, John Matthews, boatswain, recorded the extent to which they interacted with the crew and officers of the *Plover*:

> I am happy to say the natives now appear to be friendly with us and we give them all the encouragement we can, giving them dances on board. We dress and light up our ship inside and allow them all, men, women and children, to come in, and I think at these times we have all the natives belonging to the village on board, to the number of 120, or more.[133]

Expedition leaders of the searches in general felt that it was important to maintain the goodwill of Indigenous peoples, who often travelled long distances when hunting, so that they would be more disposed to help members of the lost expedition should they encounter them. With the proximity of Chukchi, Yup'ik, and Iñupiat settlements a feature of winter quarters in the Bering Strait region, there may have been less need for the distractions of an on-board periodical. For these expeditions, which spent more time engaging with Indigenous communities, there was less time available to engage in cultural production.

However, for those ships where periodicals were more popular, there was a tension between the idea of private and public as far as contributors were concerned. Blum suggests that non-contributors to Parry's endeavour may have feared for their reputations, while contributors felt what was written would stay within their private Arctic society.[134] But the practice of the subsequent publishing of the periodicals in Britain, and therefore potentially exposing contributors, also occurred. The 'New Georgia Gazette' was published as the *North Georgia Gazette, and Winter Chronicle* in Parry's narrative (1821), with some pieces that showed crew tensions excised.[135] In this case, the decision to publish was not necessarily made by contributors themselves. Two of the Franklin search productions were also published; the 'Illustrated Arctic News' and 'Aurora Borealis' were both transformed into books on the return of the Austin expedition. The publication of the latter saw a 'collection of articles . . . extracted from a newspaper' produced as *Arctic Miscellanies* (1852).[136] A handbill advertising *Arctic Miscellanies* hoped that 'such a volume will meet with a cordial reception from the public, while there can be no doubt that the Officers and Seamen of the Navy, so much interested in its achievements, will eagerly avail themselves of the opportunity of securing so agreeable a souvenir of the Search in the Polar Seas'.[137] Reviews of the *Illustrated Arctic News* were favourable, with the *Era* commenting that the expedition members' 'handy work' was 'not only peculiarly interesting but remarkably clever'. Furthermore, it revealed 'you learn here more of what transpires among our brave countrymen when they are near the North Pole than a hundred printed books can tell you'.[138] The *Literary Gazette* considered it to be 'an "Illustrated Punch", as well as an "Illustrated News", containing songs and witticisms of no inconsiderable merit'.[139] *Lloyd's Weekly Newspaper* referred to *Arctic Miscellanies* as a 'clever little production'.[140] The appearance of two of the on-board periodicals in the metropolitan sphere, while significant, was still dwarfed by the sheer amount of other publications and exhibitions – the published narratives, moving and fixed panoramas, lithographs, newspaper articles, and engravings – that told a different story. It was no wonder that the review of the *Illustrated Arctic News* in the *Literary Gazette* declared with surprise 'who would have thought that so much fun and jollity prevailed in the midst of the frozen regions on this occasion!'.[141]

As evidence of the clear difference between the private, yet collective, space of an Arctic ship in winter and the metropolitan public, a note from Osborn, written on 30 September 1854 at the front of the 'Queen's Illuminated Magazine', warned that: 'This paper must not be published or circulated publicly without the full consent of the contributors'.[142] This implies that the periodical was, in the first instance, meant for the private society on board ship and, on their return, could be viewed by select individuals. In 1857, John Barrow inscribed a note below Osborn's, outlining that Osborn had given him this 'book', which was the original 'and not hitherto printed or published'.[143]

If the 'Queen's Illuminated Magazine' was essentially a private document, intended to be read on the ship, it was also an object that held interest beyond its initial purpose of amusing the expedition members during their Arctic voyage. At the beginning of the enterprise, May signalled his own aspirations for the magazine: 'I sincerely hope that the paper will be carried out with a good will and be an amusement to us now & something to look at hereafter – that will remind us of old Arctic days.'[144] May here projected a future for the production, as a memento that they could regard long after the expedition returned. His words, 'that will remind us of old Arctic days', suggest an affection for the Arctic and perhaps even an expectation of feelings of loss in the future. In this way, the periodical was expected to serve an important purpose as a type of memento or souvenir. This would indicate that many expedition members were emotionally invested in the Arctic experience of wintering over. The lack of any marginalia on the pages of the 'Queen's Illuminated Magazine' and its good condition suggest its status as a revered object, even from the time of its making.[145]

Additional souvenirs are suggested by May's other comments regarding the forthcoming magazine: 'We published our prospectus for the Victoria Gazette[146] by the aid of the printing press – and 80 copies were struck off so that the officers should have two each and the men one copy each.'[147] The imprints of the prospectus were produced, not only for immediate consumption aboard ship – after all only a few copies would be needed for that – but expressly so that the men and officers could have souvenirs of their Arctic voyage, again indicating the emotional attachment to their experience. Other instances of this production of numerous copies of imprints, particularly dance announcements and playbills, suggest that the production of such souvenirs on other search ships was common. There is an awareness that they were taking part in an important event and that the 'old Arctic days' would inspire nostalgia and sentimentality in the future personal lives of the expedition members. Indeed, within the archived personal journals and papers of expedition members one often finds playbills and printed announcements inserted between the leaves. In *Arctic Miscellanies*, an article on their printing reveals the importance of these tokens: 'In the years to come, every little souvenir of our sojourn here will be prized for the recollections it will give rise to – of the comfort and amity that existed among the members of the "Austin Happy Family".'[148] It is notable in this quotation, too, that the period of winter 'imprisonment' is referred to as a 'sojourn', a descriptor that indicates relative rest as opposed to an ordeal.

The concept of the personal memento intended for private consumption, whether in the form of a periodical or a model sledge,[149] and its connections to memory and nostalgia is a thread that runs through many of the expeditions. Thus, periodicals, particularly those that were illustrated, as May's comments suggest, were produced not only to relieve boredom but also for their potential to be viewed in the future, as a reminder of their winter in the Arctic and the spirit of those on board. In this way, the souvenirs are very different from the idea of one that is representative of 'the exotic transported across the world'.[150] These are more intimate and meaningful souvenirs of the expedition members' own making and representative of their lived experience in the Arctic.

Conclusion

The final issue of the 'Queen's Illuminated Magazine' commented on its 'overflowing Editor's box' as the 'best proof of success of an Arctic Newspaper'.[151] Over the course of the winter, that box attracted some contributions that were never included in the magazine on account of 'extreme raciness!'.[152] Intriguingly, the editor's box also produced 'an original Operetta! and an original Pantomime! and both of great merit',[153] all suggestive of the intense cultural production that was possible on isolated overwintering expeditions.

These cultural resources are best exemplified by the on-board newspapers they produced. The use of the Arctic environment as a trigger for humour and the focus of illustrations on the domestic and familiar contrasts with the idea of the Arctic as a testing ground for British masculinity. Although partly modelled on comic magazines like *Punch* or *Cruikshank's Table-Book*, the Arctic newspapers were particular for the subjects that they chose as the targets of their humour: the cold, the aurora, imperialism, polar bears, and the expedition members themselves. In opposition to the imaginary of the Arctic as a barren wasteland, it can be seen to be a fruitful source for humorous exploitation, and the situation of the crew is replete with comic possibilities, thereby negating the Arctic sublime (whether that effect was intentional or not).

This chapter has presented an alternative narrative to the sublime Arctic by looking in detail at two illustrated periodicals during the Arctic winter. Examining the periodicals and related material reveals a strong account of familiarity with the Arctic environment and of appreciating the winter as a time of rest and of cultural production. The 'prison', so often mentioned in contemporary and modern publications, could equally be a 'sojourn', providing, in the revealing words of Osborn, 'breathing time'.[154] The need to include a supplement with the third issue of the 'Queen's Illuminated Magazine' shows the enthusiasm of contributors and the large quantity of submissions. The periodicals had other important functions in addition to providing amusement for expedition members during the dark winter. Along with some of the other material created on the expeditions, the periodicals were thought of as future mementos, expressing a desire on the part of expedition members to remain connected to each other and to their Arctic experience. Crucially, the winter period on an Arctic search ship facilitated cultural practices like periodical production. Furthermore, periodicals were most successful on ships that were completely isolated from Indigenous communities during the winter. When a large group of ships wintered in the same location, a local competitive element came into play.

The extensive media attention and the production of texts and pictures in Britain during the period of the searches was fuelled by the growing mystery of the loss of the Franklin expedition that captured the public imagination. Cultural products associated with the search, from expedition journals to sketches, had the potential to be transformed into saleable commodities – narratives, lithographs, panoramas – once an expedition returned to Britain. However, the more positive aspects of Arctic life were only partially viewed by the wider public. The following three chapters of the book all focus on the material that was produced in the public domain: narratives, panoramas, and lithographs, as well as representations in metropolitan periodicals.

CHAPTER 3

'These Dread Shores'

Visualising the Arctic for Readers (1850–1860)

In April 1819, the writer Hester Lynch Piozzi (1740–1821) penned a letter to a friend encouraging him to acquire a copy of a newly published exploration narrative. She enthused: 'But you should really get Ross's Narrative of the Arctic Expedition. One view of the Crimson Cliffs and new-discovered natives in the fine Plates annexed to that Volume; give you far better and more distinct Notions of what they saw than any words which can be put together.'[1] Piozzi was referring to the fold-out colour plates in John Ross's *Voyage of Discovery* (1819), one of the first nineteenth-century narratives of Arctic exploration. Her recognition of the superior power of the visual to convey the appearance of little-known environments lends weight to the importance of interrogating illustrations as essential components of text, acknowledging the great influence they held over the nineteenth-century eye. In Piozzi's words, 'what they saw' reveals the power of such narratives, particularly in their illustrations, to be regarded as eyewitness accounts. It also reflects the wide audience for Arctic exploration literature, as attractive to an elderly woman as it was to a naval captain.

Only ten such narratives were to be published in the first half of the nineteenth century, but the intensity of the search for Franklin saw an explosion in Arctic accounts; between 1850 and 1860, at least twenty-four Arctic exploration and travel narratives were published in book form by members of Franklin search expeditions.[2] This represented a sudden increase in books about Arctic travel, fuelled by the large number of expeditions (thirty-six) that took part in the search and the public interest in Franklin's mysterious disappearance. This chapter deals with the representation of the Arctic in these books, specifically with those produced by maritime (as opposed to overland) voyages, and with the mediation of both pictures and texts through their publication. The majority of these Arctic narratives are illustrated, ranging from a single tinted frontispiece in Robert Goodsir's *An Arctic Voyage to Baffin's Bay* (1850) to over three hundred monochrome prints in Elisha Kent Kane's *Arctic Explorations*

(1856). The importance of the 'dialogue between word and image' in illustrated books requires the use of both 'visual and verbal interpretation, in order to show how the interaction between pictures and words produces meanings'.[3] The 'visualising' of the Arctic, in both senses – through actual printed plates and through descriptions in words that invoke images, or phantasmata, in the mind of the reader – is of key importance here, demonstrating how a very visual Arctic was created for readers. The representation of the Arctic was heavily dependent on factors such as the success and duration of the voyage, the way in which an author wished to portray himself,[4] and the visual culture of the nineteenth century.

The authors and illustrators of narratives drew on popular visual culture to aid their representation of the Arctic. The influences of panoramas, paintings, exhibitions such as *tableaux*, and magic-lantern shows can be discerned in the choice of language and illustration. As well as this, the technique of virtual witnessing, Romantic poetry, and referencing experiential memories all served to make the Arctic more visible for readers. The illustrations themselves can often be seen to display a triadic canon of motifs that became associated with Arctic exploration, ensuring that the main themes of the story are evident by merely glancing at the illustrations: masculinity signified by hunting; peril signified by the ship trapped in the ice; and exploration signified by new horizons and headlands.

First-time visitors to the Arctic, particularly those on ill-prepared expeditions, employed aesthetic language to describe the environment they encountered. The 'man versus nature' trope, so often used in the narratives of polar exploration, appears with regularity throughout texts. Integral to this was the setting of the narrative and the stark geology and ice of the Arctic that can be seen to resent or 'frown' upon the intruders who dared to approach its 'dread shores'.[5] By contrast, the narratives of repeat visitors to the Arctic, and those who spent consecutive winters there successfully, show a decrease in their use of aesthetic language as they become embedded in the environment as a local place, in practical terms or in its emotional resonance acquired through memory. This was particularly facilitated by the act of wintering over, a method of securing a home in the ice that was possible on ship-based maritime expeditions. In this way, some narrators of the Franklin search expeditions differ in their mode of looking from British explorers described by Mary Louise Pratt, who search for the source of the Nile with the 'monarch-of-all-I-survey'[6] mode of looking. For these long-term visitors to the Arctic, the environment was viewed in more practical, less aesthetic terms and the surrounds of winter quarters became their locality, causing them to lean towards behaving more like 'inhabitants' for whom the same spaces are 'lived as intensely humanized, saturated with local history and meaning'.[7] Indeed, the establishment of winter quarters enabled the creation of places, or 'spaces people are attached to in one way or another ... a meaningful location'.[8] Yi-Fu Tuan equates space with movement and place with pauses,[9] reminding us that the immobility of the Arctic search ship in winter is distinct from the ship that moves through the environment.

However, the printed visual depictions in such narratives often jar with the verbal picture, showing only the Arctic in winter and failing to display the unthreatening greenery and foliage of the Arctic summer that is described in the text. Such representations display the influence of the publisher's commercial interests and the hand of the metropolitan artist. This mirrors visual representation of the region today where an overemphasis on winter has led to the general perception of the Arctic as a frozen icescape; twenty-first-century narratives, such as that surrounding the Greenland ice-core drillings, reproduce the Arctic as an 'empty, frozen space, waiting to be conquered by scientists'.[10] This attitude continues to configure the Arctic as the testing ground for white masculinity, an image that is not compatible with flora and greenery. In contrast, the Inuit climate change, cultural, and human rights advocate Sheila Watt-Cloutier references the 'preconceived notions of the Arctic and Inuit that many people hold' when she tells us as clearly as possible: 'The Arctic is not a frozen wasteland.'[11]

Although critics such as Jen Hill have pointed out that in the nineteenth-century popular imagination the Arctic was empty,[12] narratives written by members of search expeditions can be full of Indigenous presences when they wintered near local settlements. Furthermore, the ways in which Inuit, Chukchi, and Iñupiat were represented in the narratives by those who had long-term contact with them suggest that certain individuals and groups were held in higher regard than the lowest rank on board ship, that of the seaman. After all, expeditions used them for navigation, clothing production, hunting, and social interaction, often benefiting from an astounding level of hospitality and goodwill on the part of Indigenous families. However, it can be said of both seamen and Indigenous Arctic peoples that the attitudes of the higher-ranking British expedition members towards them were paternalistic, perhaps in many ways similar to the approach of Danish colonialism in Greenland, which Nanna Katrine Lüders Kaalund identifies in the narrative of Carl Petersen, whose 'approach to asserting his superiority to the Inuit' was 'shrouded in paternalistic rhetoric'.[13]

Between 1850 and 1852, seven travel narratives by members of Franklin search expeditions were published. In 1853, a further five Arctic travel narratives of the search appeared in print. Unlike the Arctic exploration narratives from 1818 to 1848, which were (largely) published by John Murray and controlled by John Barrow (First Secretary to the Admiralty),[14] the narratives of the mid-nineteenth century were made available by a wide range of companies, including the firm of Longman, Brown, Green, and Longmans, which published four Arctic narratives between 1851 and 1852. The cost of narratives varied dramatically during the period, from one shilling and sixpence for a small single volume to thirty-six shillings for a heavily illustrated two-volume narrative.[15] As Adriana Craciun points out, the death of John Murray II in 1843 and of John Barrow in 1848 ended the 'polar print nexus' and 'made possible the explosion in popular Victorian writings on the Arctic'.[16] Of course, this new explosion in writing was also made possible by the large number of expeditions participating in the search for Franklin.

This chapter looks at the representation of the Arctic in a broad range of maritime narratives from the search expeditions, including those of Robert Goodsir, William Parker Snow, and Edward Augustus Inglefield, who are all distinguished by writing narratives of summer voyages. In contrast, I also look at the narratives of those who spent winter in the Arctic, like Sherard Osborn. Given the prominence of Osborn in the history of exploration literature of the Arctic, not least as the editor of two on-board periodicals, I pay a significant amount of attention to his published narrative.[17] Furthermore, repeat visitors like Francis Leopold McClintock, who voyaged to the Arctic four times before (reluctantly) publishing a narrative, and George Frederick McDougall, who wrote a narrative of his second expedition to the Arctic, are important in showing how the Arctic became familiar. Another type of representation is evident in that of the long-term visitor like William Hulme Hooper, who spent several consecutive winters in the Arctic. The narratives of Isaac Israel Hayes and Elisha Kent Kane, who were on the ill-equipped Grinnell expeditions, represent the disastrous voyages for which the Arctic has become renowned.

Techniques of Making the Arctic Visible

From the late eighteenth century to the 1850s, 'geographical exploration and travel tales captivated public audiences and journalistic commentators alike. Everywhere, it seemed, was coming under the explorer's gaze.'[18]

Explorers' accounts often achieved bestseller status, providing excitement for readers with the authenticity of a real adventure.[19] Texts like Constantine Phipps's *A Voyage towards the North Pole* (1774) and Samuel Hearne's *Journey . . . to the Northern Ocean* (1795) provided later authors of both literary and exploration narratives with models on which to construct their Arctic imagery. Early nineteenth-century Arctic exploration, which had resulted in numerous publications and exhibitions, meant that the Arctic was firmly established in the public imagination when Franklin set out on his final voyage in 1845.[20] Narratives dealing with Arctic exploration, like William Edward Parry's *Journal of a Voyage for the Discovery of a North-West Passage* (1821) and Ross's *Voyage of Discovery* (1819), were presumably well read on board the search vessels. We are fortunate in having access to a library catalogue[21] that was printed aboard the *Assistance* during the Belcher expedition of 1852 to 1854. This records at least sixty travel and exploration narratives in the collection, including Parry's and Franklin's from the 1820s, Edward Belcher's Pacific narratives, and *Two Years before the Mast* (1840), a memoir describing life as a common seaman by the American Richard Henry Dana, which had been published in Britain in 1845.[22] This presence of travel and exploration narratives, representing at least 9 percent of the library's total,[23] meant that expedition members were well aware of the genre and attuned to the publication potential of their own experience. Books on geography, history, travel, and biography made up 40 percent of the library's collection.

Through their use of language and illustrations, the authors and illustrators of narratives drew on popular visual culture (particularly the nineteenth-century culture of exhibition) and memory (both personal and collective) to render the Arctic visible for readers. Visual media also strongly affected the way in which a first-time expedition member, like William Parker Snow, witnessed the Arctic. With illustrations, poems, and vivid verbal descriptions all leading towards creating a visual Arctic for the reader, the technique of 'virtual witnessing' was also incorporated into the narrative, whereby authors generated naturalistic images in the minds of readers through the use of language and by their choice of illustrations.[24] These visualisations help to form memories and phantasmata, or images in the mind.[25]

In Osborn's *Stray Leaves*, the third plate, *Winter Quarters*, is placed beside the corresponding text on the facing page.[26] The illustration is peaceful, with the four ships represented against a lightening sky and bright icescape. The horizontal lines of the ice and the topography create a calm atmosphere and our eye is drawn to all the interesting details in the lower third of the picture, under an immense Arctic dawn. The scene connects with Osborn's invitation to the reader to enter into his virtual panorama, reminiscent of the experience of viewing a panorama in London. Initially, Osborn described the environment as the expedition members experienced it: 'a full, silvery moon . . . threw a poetry over every thing, which reached and glowed in the heart, in spite of intense frost and biting breeze. At such a time we were wont to pull on our warm jackets and seal-skin caps.'[27] This 'striding out upon the floe . . . under so bracing a climate'[28] and looking down on the 'squadron' from the 'heights of Griffith's Island'[29] reveals the experiential element of the sojourners' stay in winter quarters. The practicalities described – clothing, physical movement through the environment, and emotions – are in contrast to the panorama that Osborn then revisited over forty-four lines, like a guide, for the benefit of the reader:

> Imagine yourself, dear reader, on the edge of a lofty table-land . . . fancy a vast plain of ice and snow, diversified by snow-wreaths, which, glistening on the one side, reflected back the moonlight with an exceeding brilliancy, whilst the strong shadow on the farther side of the masses threw them out in strong relief. Four lone barks, atoms in the extensive landscape, – the observers' home, – and beyond them, on the horizon, sweeping in many a bay, valley, and headland, the coast of Cornwallis Island, now bursting upon the eye in startling distinctness, then receding into shadow and gloom, and then anon diversified with flickering shades, like an autumnal landscape in our own dear land, as the fleecy clouds sailed slowly across the moon, – she the while riding through a heaven of deepest blue . . . and say if the North has not its charms for him who can appreciate such novel aspects of nature.[30]

Osborn directs the reader where to look and what to look at, like a tour guide in a virtual landscape. In this way, the reader is encircled by an imaginary fixed panorama, but one that includes 'home' for these particular observers. It is probable that Osborn is here playing on his readers' familiarity with Robert Burford's popular fixed panorama, *Summer and Winter Views of the Polar Regions* (discussed in Chapter 4), which was exhibited in London in 1850 and was based on sketches by Lieutenant William Henry Browne from the Ross expedition of 1848 to 1849.

William Hulme Hooper, in *Ten Months among the Tents of the Tuski* (1853), also invited the reader into his book to share 'one of the most exhilarating incidents one can experience': 'We have just arrived at the head of a bad rapid, and are preparing to "run" it: will you take a seat with us in the boat, reader, and share the striking episode? Come then.'[31]

Hooper addresses the individual reader directly in the present tense with a sense of immediacy; he invites us to sit beside him in the canoe in the Mackenzie River district with the 'water bubbling and foaming and roaring around us, spray dashing into our faces'.[32] Hooper keeps the reader in the action by periodically urging us to observe: 'but see, see!'; 'But look, look!'[33] Hooper has gone further than Osborn; his use of the imperative heightens the sense of the occasion as he instructs the reader. Osborn wanted us to imagine a visual experience without the impracticalities of Arctic life; Hooper asked us to take a seat and, over the course of four pages, littered with exclamation marks, the reader imagines feeling physically wet. Osborn's is visual; Hooper's requires imagined participation.

In essence, these mid-nineteenth-century authors go further than just picturing the landscape; they exhibit it and invite the reader into an imagined experience. While the authors of these narratives employ visuality, beyond mere landscape description, to engage their readers, first-time visitors and summer visitors to the Arctic often referenced popular visual culture directly to describe sights. Isaac Israel Hayes, in his narrative *An Arctic Boat Journey in the Autumn of 1854* (1867), invited the reader into the explorer's world, via a scene presented as an instructional and ethnological exhibit: 'If the reader will follow me into the hut he will see there a succession of *tableaux* which may be novel to him. The two above-mentioned hunters sit facing each other, and facing the lump of frozen beef, which lies upon the ground.'[34] When William Parker Snow witnessed Port Leopold, where the Ross expedition had spent the winter of 1848 to 1849, he drew direct parallels with his experience of viewing the panorama *Summer and Winter Views of the Polar Regions* in London earlier that year:

> As we neared the shore, the whole features of the place came fresh upon me, so truthful is the representation given of them by Lieut. Browne in Burford's panorama. I could not mistake; and, I, almost, fancied that I was again in London viewing the artistic sketch, but for certain undeniable facts in the temperature and aspect of the ice which banished such an idea.[35]

The interplay here between the represented landscape and the experience of being in the Arctic suggests that the author has been conditioned by visual culture in such a way that he does not privilege the authentic experience. While viewers of the panorama in London could imagine themselves in the Arctic, here, Snow reversed that dynamic, re-imagining himself back in London at the panorama. The panorama was Snow's primary source, and he projected the Arctic of Burford onto the world he encountered, proving that travellers sometimes see what they expect.[36]

Humboldt recognised this phenomenon when he noted that spectators of panoramas were immersed in the experience: 'Impressions are thus produced which in some cases mingle years afterwards by a wonderful illusion with the remembrances of natural scenes actually beheld.'[37]

As well as presenting the Arctic and its people as an exhibit, authors also regularly turned to Romantic poetry to describe the nature they encountered, and texts were embedded in a web of literary references. Here, Osborn was not content with conveying an exciting scene with crisp active verbs and breathless punctuation for the reader. He felt it necessary to romanticise the novelty of the ice by using the words from Coleridge's 'Rime of the Ancient Mariner' to communicate his own personal experience:

> *June 24th, Baffin's Bay.* – The squadron was flying north, in an open sea, over which bergs of every size and shape floated in wild magnificence. The excitement, as we dashed through the storm, in steering clear of them, was delightful from its novelty. Hard a starboard! Steady! Port! Port! you may! – and we flew past some huge mass, over which the green seas were fruitlessly trying to dash themselves. Coleridge describes the scene around us too well for me to degrade it with my prose. I will give his version: –
>
> 'And now there came both mist and snow,
> And it grew wondrous cold,
> And ice, mast high, came floating by
> As green as emerald.'[38]

Hooper, too, recalled the 'Rime of the Ancient Mariner' when their journey was halted by ice: 'We had a full view of the impossibility of a present advance, – "The ice was here, the ice was there, / The ice was all around."'[39] In fact, Coleridge had based his visualisation of ice on earlier narratives of exploration and the perils they could entail.[40] In particular, the influence of James Cook's second voyage, in which the *Resolution* was the first ship to cross the Antarctic Circle in 1773, on Coleridge's poem was highly likely, as William Wales, the expedition astronomer and meteorologist, taught at Christ's Hospital School during the time in which Coleridge was there.[41]

Finally, referencing their own memories of England to aid their description of landscapes was not an uncommon feature of Arctic narratives. It has been argued that the appropriation of the English landscape into the Arctic by British naval officers in the nineteenth century betrayed a lack of comprehension of the Arctic environment, which could even lead to disaster.[42] Osborn was no exception, and he did indeed use the English countryside to describe the scene to his readers. Near Melville's Monument, on the west coast of northern Greenland, he used the English landscape as a simile: 'a firm unbroken sheet of ice extended to the land, some fifteen miles distant. Across it, in various directions, like hedge-rows in an English landscape, ran long lines of piled-up hummocks, formed during the winter by some great pressure.'[43] However, it is equally possible that, in this case, Osborn did so for the benefit of a readership that would never travel to the Arctic, rather than to aid his own comprehension of the landscape. Ann Radcliffe defended her use of this technique in her account *A Journey Made in the Summer of 1794*, when she clarified that 'one of the best modes of describing to any class of readers what they may not know, is by comparing it with what they do'.[44]

Other environmental and experiential memories that appear repeatedly in British Arctic narratives, perhaps surprisingly, are those of exotic places far from the shores of both Britain and the Arctic. The tropics, where many of the Arctic officers had recently served on surveying missions, provided a surprisingly useful, often contrasting, reference point.[45] In *Ten Months among the Tents of the Tuski*, Hooper realised that Arctic nights can be as 'charming', although very different, as those in hot climes:

> The harbour is freezing over fast, with the water as smooth as glass: a bright moon and cloudless starlit sky render the scene one of the most perfect for tranquil beauty I can remember ever to have witnessed; yet here are no trees, no woods, no foliage to enliven the view; – all is snow-clad; mountains and rugged hills frowning in their majesty where thrown into deep shade, and assuming with the headlands and slopes strange fantastic shadows, jutting out in bold relief on the silent water. So, after all, there may be other than tropic nights charming.[46]

Like other narrators who had been employed in the Pacific, Osborn used the Arctic to imagine its antithesis, by connecting memory and sentiment. While observing the geology of the limestone cliffs of Devon Island with their 'numerous fossil shells, Crustacea, and corallines', he and his companions were sentimentally transported 'back to the sun-blest climes, where the blue Pacific lashes the coral-guarded isles of sweet Otaheite, and I must plead guilty to a recreant sigh for past recollections and dear friends, all summoned up by the contemplation of a fragment of fossil-coral'.[47]

The Environment Made Visible: Icescapes and Landscapes

The Arctic environment 'placed on view' prompted a wide variety of published aesthetic responses from the expedition members in the years of the Franklin searches. These reactions, transmitted through the descriptive passages and illustrations, were tailored for their expected readership. The language employed and illustrations printed made the Arctic visible for readers of the narratives. Never before had so many areas of the Arctic, from the Chukchi Peninsula to the southern tip of Greenland, been described, denounced, or admired. Not all regions of the Arctic were aesthetically pleasing to the nineteenth-century eye or deemed suitable for illustration. A lack of visual novelty was associated with the western reaches of the passage, for example. Although there is considerable climatic and topographical variation, the public perception of the Arctic was as 'a dangerous zone where visitors risked their lives'.[48] The map (Figure 3.1) shows the geographical spread of the regions described in the narratives. Some voyages produced narratives by two authors, and it can be seen that 70 percent of all published narratives during this decade covered the Atlantic approach to the archipelago via Baffin Bay and Greenland, which corresponds to the approximately 70 percent of search ships that approached from the Atlantic.

The immense icebergs of Baffin Bay, in particular, calved from Greenlandic glaciers,[49] elicited wide ranges of responses varying from disgust to delight. Despite the fact that numerous whaling ships visited Baffin Bay during summers in the nineteenth century, this region is visualised almost as another world in the search narratives. Robert Goodsir, who sailed to the Arctic on a summer whaling cruise with Captain William Penny in 1847, alluded to Edmund Spenser's late sixteenth-century poem *The Faerie Queene* when he enthused: 'it is impossible to describe the beauties of these ice islands ... The imagination of Poet or Painter never fancied grotto fitter for a Fairy Queen than these would be, could but the beauties of the Floral world be associated with them.'[50] Goodsir saw a landscape both picturesque and medieval, his ice grottos replacing the ruined castles so popular in the picturesque mode.

For William Parker Snow, on the first Arctic voyage of the *Prince Albert* in 1850, a delicate icescape unfurled like a panoramic fairy tale. His published narrative, *Voyage of the Prince Albert* (1851), based on a summer search, was illustrated with four soft, delicate colour lithographs. *The 'Prince Albert' Surrounded by Icebergs* (Figure 3.2) resounds with the more positive and fairy-tale qualities of Snow's text and shows a dream-like icescape, likely based on an image constructed after the voyage and incorporating a large proportion of his imagination. The closely clustered ice of the background, although pointed and sharp, is rendered in soft, pastel

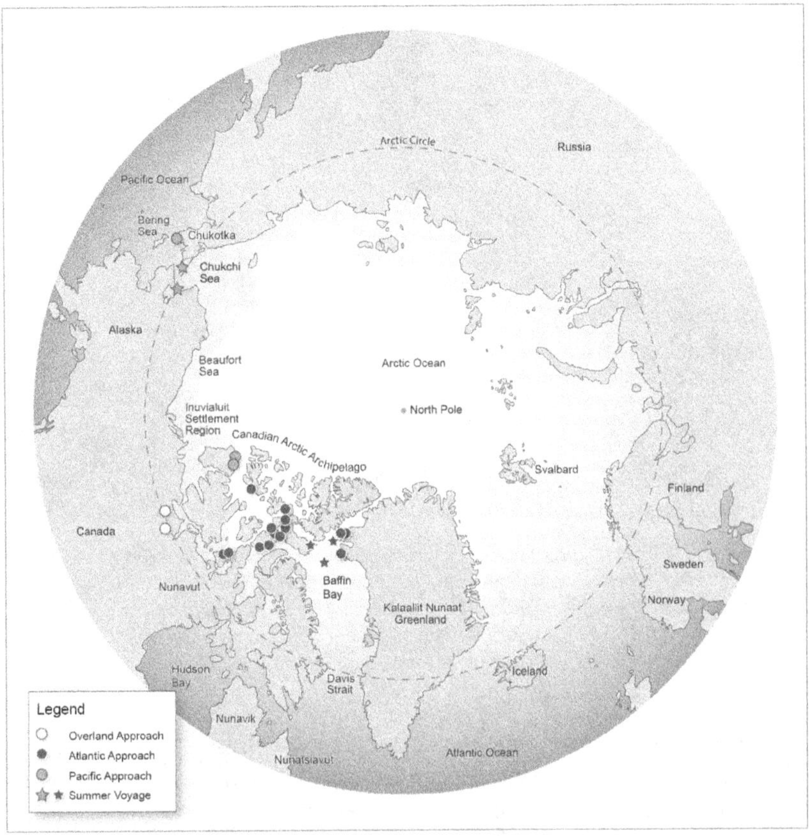

Figure 3.1 Main locations of published narratives, 1850–60, from the search and supply expeditions. (Locations refer to areas in which ships spent most of their time.) Map by author.

tones, creating a nebulous appearance that jars with the potentially dangerous situation. This softness reflects the author's delight in the Arctic icescapes as expressed in the textual description. However, the seductive appearance of the ice could be seen to have parallels with fairy tales in which danger and supernatural power is disguised as something ambiguous, like the gingerbread house in the woods.[51]

In contrast, Commander Edward Augustus Inglefield, whose narrative related a summer in the Arctic on the small screw schooner[52] the *Isabel* in 1852, described the icebergs of Baffin Bay as 'rude and grotesque'.[53] Indeed, they become personified, 'like "grim watchmen" of the bay, scowling on

Figure 3.2 William Parker Snow, *The 'Prince Albert' Surrounded by Icebergs, Voyage of the Prince Albert*, 1851. Colour lithograph, 12.5 × 20 cm. Frontispiece of Snow, *Voyage of the Prince Albert in Search of Sir John Franklin: A Narrative of Every-Day Life in the Arctic Seas* (London: Longman, Brown, Green, and Longmans, 1851). Courtesy of Toronto Public Library.

the adventurous stranger who would tempt these dread shores' in the sea off Cape York, on the west coast of Greenland.[54] This negative personification of the Arctic landscape repeatedly occurs in some narratives, leading to an image of Arctic geology as a harsh entity that eternally 'frowns' at readers.[55] This kind of anthropomorphising of rock occurs in gothic novels like Ann Radcliffe's *Mysteries of Udolpho* (1794), when the female protagonist, Emily, sees the villain Montoni's castle for the first time: 'Silent, lonely, and sublime, it seemed to stand the sovereign of the scene, and to frown defiance on all who dared to invade its solitary reign.'[56] Inglefield's terse descriptions of the Arctic are presented to the reader with the implication that his summer voyage was in fact as perilous as expeditions that spent longer periods in the Arctic. His emphasis on new discoveries, on-the-spot sketching, and inaccuracies in previous charting suggests his need to justify his endeavour and this publication.[57] Keeping his narrative formal and stressing scientific accuracy, Inglefield cannot resist the language of the negative sublime, creating a malevolent visual image of the Arctic that both attracted and repelled. Ultimately, he writes himself as the heroic explorer against the 'dread' Arctic, aligning himself with a polar sublime and 'partaking of the sublimity against which he matched himself'.[58]

Although the narrative contains only five illustrations, the pictures chosen for print tell an independent story by using the triadic canon of motifs to highlight both danger and discovery. The illustrations *Killing a Bear off Cape York* and *Dangerous Position of the 'Isabel' Caught in the Lee Pack* serve to emphasise the masculine activity and danger associated with Arctic voyages; the visual image of a leaning ship caught in pack ice is a ubiquitous motif in Arctic publications, signifying peril and the possibility of disaster.[59] However, the calm and tranquil fold-out panoramic print that illustrates Inglefield's apparent discovery of the entrance to the 'Open Polar Sea', *Midnight Aug 26th 1852*, as well as the frontispiece, *'Isabel' Entering the Polar Sea through Smith's Sound, Midnight*, show a benign Arctic, suggesting an entrance to a hidden paradise:[60] 'We were entering the Polar Sea ... plying onward in the unfreezing Polar Basin.'[61] Inglefield was, of course, completely mistaken, and the theory of an unfrozen sea surrounding the North Pole was later invalidated.

First-time visitors who overwintered on ill-prepared expeditions also created extreme visualisations of the Arctic environment, attributing an almost supernatural quality to a geographical region. As Craciun shows, with regard to Franklin's account of his disastrous Arctic land expedition, *Narrative of a Journey to the Shores of the Polar Sea* (1823), distressing ordeals had the potential to become narratives and then commodities.[62] A narrative that included 'starvation, murder, cannibalism, and madness was tailored to please an audience already schooled in Gothic romances, captivity narratives, shipwreck accounts, early ethnography, and travel writings.'[63] The American Grinnell expeditions, unprepared and badly planned,[64] were already best-selling narratives waiting to happen. Elisha Kent Kane's *Arctic Explorations* (1856) was praised by a reviewer as a demonstration of human endurance; the 'terrible record of suffering' was 'almost too horrible to read'.[65] Isaac Israel Hayes, on that second Grinnell expedition led by Kane, imagined his 'home-world' when trying to escape the Arctic, suggesting that for him the Arctic was another world: 'Far behind that dreary mist lay our home-world, gladdened by a genial sun – glowing in the gold and crimson of its autumn. The pictures which our fancy drew made such contrast with the realities of our situation, that we fell to scheming again for our deliverance.'[66] Hayes's imprisonment in the Arctic, unable to escape, echoes the captive trope of Gothic novels, lending the appeal of excitement to the respectability of an expedition narrative, all the more thrilling for its basis in reality. Although Hayes used the term 'home' to refer to their temporary hut throughout the text, his 'home-world' is the place to which he yearned to return. His use of the term 'dreary' is typical of Arctic narratives; it was a common adjective that had become associated with landscape and particularly with the landscape of the North since the eighteenth century, for example, in James Beattie's *The Minstrel* (1771), 'the chill Lapponian's dreary land' connects the adjective to winter in Lapland.[67] The strong connection of the word 'dreary' with the Arctic was compounded by the Romantic poet and political commentator Robert Southey, whose *Life of Nelson* (1813) was included in the library of the *Assistance* during the Belcher expedition.[68] As a young boy, Nelson had participated in Phipps's Arctic voyage to Spitsbergen (now Svalbard), and, although Southey drew on Phipps's narrative for his description of the environment, the former introduced descriptors such as 'dreary' to describe the Arctic in the *Life of Nelson*. This influential and imperialist book was 'one of the texts by which Romanticism educated the British in an ideology of empire'.[69] However, it is notable that the term 'dreary' is not one that is used in Phipps's original book.[70]

By contrast, authors who participated in relatively well-organised expeditions and spent several winters in the Arctic became inured to it and felt more at ease in the environment.[71] Their winter 'imprisonment' in the ice-bound ship reads more like a retreat from civilisation, in fact they often use the word 'sojourn',[72] implying a restful stay and aligning with the idea of travel as an escape from civilisation.[73] Lengthy descriptions of landscapes lessen when the explorer is embedded in the environment. For example, McClintock participated in four northern voyages between 1848 and 1859, experiencing a total of six winters in Baffin Bay and the Canadian archipelago.[74] For this reluctant author, aided by Osborn who '[saw] it through the press', 'lofty' is the only term in his book that approaches a romantic aesthetic when the mountains of Greenland or Baffin Island appear.[75] This stock romantic adjective produces images of high mountain peaks, wreathed in mist in the mind of the reader. It is likely that the cause of McClintock's apparent lack of 'aesthetic interest in nature'[76] was his familiarity with the region, having spent much of the 1850s between Greenland and the Canadian archipelago.

Furthermore, the knowledge of the topography of local areas where ships had wintered caused the landscape to be experienced with new meaning and even with affection by those returning to the region. McDougall, returning to the Arctic for the second time on the Kellett arm of Belcher's squadron in 1852, noted how the landscape became familiar and now held new significance:

> The sight of Griffith's Island, as may be supposed, afforded a peculiar interest to those who, forming a part of the expedition under Captain Austin, had spent eleven months, frozen up in its immediate neighbourhood. Every point, hill, and ravine, was connected with some little incident, during our rambles over its desolate, and uninteresting surface.[77]

McDougall here eschews the popular term 'imprisonment' for the phrase 'frozen up', also favouring 'neighbourhood', a word describing a local, familiar area, a place with meaning as opposed to a landscape. Here, the landscape has become a locality. It is no longer described in pictorial terms. No textual description leads the eye from the foreground towards the horizon. Instead the island is a local place, infused with personal memories triggered by its sight. 'Every point' is connected with 'some little incident', suggesting the many small occurrences that made up the life of the expedition in winter, and the choice of the noun 'rambles' evokes a leisurely walk undertaken for pleasure, associated with a sense of freedom and childhood adventure. The place, despite its 'desolate, and uninteresting surface', is now of 'peculiar interest' because of the emotional attachment that McDougall and his fellow former members of the Austin expedition feel. The members of Horatio Austin's expedition are also bound to each other by their links to the island, marking them out as a distinct group, separate from those who were new to the neighbourhood. McDougall's inclusion of the phrase 'as may be supposed', which connects with what he imagines are his readers' suppositions regarding the sight of the island, implies that he assumed his readers would understand this intimacy with the environment.

Although places like Griffith Island may have seemed topographically 'uninteresting', published narratives needed to include eye-catching illustrations. These incorporated clear visual prompts that could transport readers pictorially to the polar regions. Inevitably, these include ice and snow in abundance. Even summer visitors focused on the more dramatic nature of the ice to seduce readers, while greenery was rarely depicted in the published illustrations. This is despite the fact that most outdoor sketching on these voyages was done between May and September, during the Arctic summer, when vegetation appeared. The visual preoccupation with ice and snow no doubt accounts for the misconception, prevalent today, that Arctic regions are perpetually snow-clad.

Elisha Kent Kane's first Arctic narrative took this to an extreme with the *Inspector's House – Lievely* depicted as covered in snow, unlikely on the west coast of Greenland in late June when Kane visited and indeed not mentioned in the text that describes his impressions of Lievely (Qeqertarsuaq) and his description of the house on subsequent pages.[78] In the picture, the snow falls heavily, and icicles hang from an adjacent structure. While the brief and undated caption leaves much to the imagination, the associated text of the chapter unsettles the illustration. Although 'snow, as usual, covered the lower slopes [of the hills]', Kane found a profusion of flowers and berries like 'a carpet pattern of rich colours' behind the settlement. He declared: 'The Arctic turf is unequalled: nothing in the tropics approaches it for specific variety'.[79] Despite the extensive use of prints in Kane's books, not one depicts Arctic flora, for example, preferring instead to highlight sharp peaks, ships leaning at acute angles, and immense icebergs, suggesting the terror associated with the sublime. The exclusively monochrome illustrations in his narrative leach any colour from the Arctic and show a landscape only in terms of dark and light. A note at the beginning of Kane's narrative explains that the author was unable to revise the book before it was published, as he departed on his second Arctic expedition in May 1853,[80] which may account for some of the jarring disconnection between text and picture.

An unusual visual representation of flora exists in the Arctic narratives of this period. It must have surprised the reader used to being presented with icebergs and snow or desolate hills. Peter Sutherland, the surgeon on the Penny expedition of 1850 to 1851, included a colour representation of Arctic flora in his narrative.[81] Although the expedition wintered near present-day Resolute, deep in the Canadian Arctic archipelago, the collection of flora is set against the backdrop of an ice-free Assistance Harbour. This comparatively expensive book (twenty-seven shillings), in two volumes, was aimed at a more scientific audience.[82] *A Group of Thirteen of the Flowers most Commonly Found around Assistance Bay* (Figure 3.3) is one of six colour lithographs used to illustrate the narrative. Although the flowers are included as 'specimens', the inclusion of their environmental context in the background associates them clearly with the Arctic.[83] The bright and fresh species of Arctic flora in bloom takes up most of the picture, while also acting as a frame for the ships, which become small and insignificant in the distance.

This unique representation of an Arctic summer, and its pockets of verdant lushness, has little illustrative parallel among the body of narratives of the Franklin search expeditions. Although textually we find references to

Figure 3.3 Peter C. Sutherland, *A Group of Thirteen of the Flowers most Commonly Found around Assistance Bay*, 1852. Colour lithograph, 17.8 × 11.6 cm. Plate 4, vol. 2 of Sutherland, *Journal of a Voyage in Baffin's Bay and Barrow Straits . . . in Search of the Missing Crews of H.M. Ships Erebus and Terror*, 2 vols. (London: Longman, Brown, Green, and Longmans, 1852). Courtesy of Toronto Public Library.

such scenes, prints of flora, small and feminine, presumably were not wholly compatible with the masculine image of the search. In his introduction, Sutherland thanked 'the artists employed upon the illustrations'. He furthermore trusted that 'the public will appreciate the arrangement and skill of the lady who has, from dried specimens, presented them with a plate, exhibiting the most abundant species of the thinly-scattered flora of Cornwallis Island'.[84] That the 'lady' is separated from the 'artists' who worked on the other illustrations corresponds to the separation of the flora from the more typical Arctic scenes in narratives and its association with femininity.

Although many authors commented on the plant life found in the Arctic, such scenes were not generally illustrated in the narratives. McClintock noted that on the west side of Baffin Bay, near the entrance to the Northwest Passage: 'Upon many well-sheltered slopes we found much rich grass. All the little plants were in full flower; some of them familiar to us at home, such as the buttercup, sorrel, and dandelion.'[85] Osborn's written descriptions also include the Arctic summer, when in July of 1850 he amused himself on the Duck Islands near Devil's Thumb by 'picking some pretty Arctic flowers, such as anemones, poppies, and saxifrage, which grew in sheltered nooks amongst the rocks'.[86] This briefly glimpsed floral vision was wholly incompatible with the icy wastes that had become associated with the polar regions in the early nineteenth century, through texts like Mary Shelley's *Frankenstein* (1818) and Radcliffe's poem 'The Snow-Fiend', published in 1826, in which a 'vengeful' being rules the North with a 'savage eye':

> From half-known Greenland's snow-piled shore
> To Newfoundland and Labrador;
> O'er solid seas, where nought is scanned
> To mark a difference from land,
> And sound itself does but explain
> The desolation of his reign.[87]

Summer visitors tended to have the most polarised reactions to the environment, as the scenery for a fairy tale or as an openly hostile landscape. These authors conveyed the most other-worldly images for their readers. Those on unsuccessful expeditions, like Hayes, were anxious to convey their perceived harshness of the environment and presented a negative Arctic sublime, heavily coded and pictorialised for the reader. Conversely, authors who spent more time in the Arctic, especially some who had relatively successful winter sojourns, portrayed it in more practical terms, and sometimes with comfortable familiarity, suggesting that they no longer viewed it as a formal landscape presented for view (or a series of pictures), but became embedded in their locality. For those authors who wintered in the Arctic repeatedly, the landscapes ceased to be framed pictorially and became localities imbued with emotional resonance through personal and experiential knowledge of the region.

Representing Indigenous Peoples

The depictions of the Arctic did not feature the physical environment alone but also, on occasion, Indigenous peoples who lived there. A major visual feature of all maritime Arctic travel narratives during the Franklin search is the image of the Northern inhabitant that was transmitted to the 'civilised' world. Travel literature helped to shape conventions of the representation of various 'races', both by the showmen who exhibited actual people in popular, and accessible, displays during the nineteenth century and by the public who interpreted the shows; travel literature provided a resource of descriptive information used to contextualise such exhibitions.[88] Ethnologists were also reliant on travel narratives for information on 'races of the world'.[89] This picturing was committed to the published record safe in the assumption that the Indigenous person would never read the book in which they were described. By the 1840s, ethnology had emerged as a discipline that used 'physical, social, and linguistic features' to study human variation.[90] James Cowles Prichard, who wrote the chapter 'Ethnology' in the *Manual of Scientific Enquiry* (1849),[91] recommended that the traveller study 'all that relates to human beings' in 'remote nations',[92] with attention to physical characteristics, society, and language, poetry, and literature. Inuit had been sensitively represented by the ethnologist, surgeon, and Arctic explorer Richard King, who published a series of three articles on their 'industrial arts', 'intellectual character', and 'physical characters' in 1848.[93] As Ellen Boucher observes, expeditions had no problem in appropriating Inuit survival techniques, but their debt to Inuit was 'easily obscured and forgotten'.[94]

Attitudes to the local populations are starkly representative of their time. In the mid-nineteenth century, the view of the Indigenous person as a 'noble savage' was still popular; in this context, 'savages' were often viewed in a sentimental manner.[95] In the published narratives, the representation of Indigenous peoples in the Arctic, or the characters they often become, is clearly influenced by merging filters of aesthetics, gender, ethnicity, and unequal power relations. While their visibility is sometimes clear, lower ranks of the expedition itself, such as seamen, are shadowy figures in the narratives, known collectively but rarely individually. Attitudes towards seamen, or 'Jack' as they were commonly known, even suggest that, in terms of class, some Indigenous individuals were held in higher regard.

There are two major regions in the Franklin maritime search area that can be thought of in terms of Mary Louise Pratt's 'contact zones', which are 'social spaces where disparate cultures meet, clash, and grapple with each other, often in highly asymmetrical relations of domination and subordination'.[96] These are the Inuit communities around Baffin Bay, with which expeditions had passing contact, and the Chukchi, Yup'ik, and Iñupiat areas around the Bering Strait, where expeditions that wintered over had prolonged contact with Indigenous communities. Other areas include the North American mainland and the southern part of the archipelago where expeditions had contact with Inuvialuit and Inuinnait (Copper Inuit). For the search expeditions that approached the Northwest Passage via Greenland and Baffin Bay, their first contact occurred at one of the harbour villages on the west coast of Greenland. The appearance of the settlements, whose residents were sometimes a mixture of Danish and Inuit parentage, often aroused disgust in the first-time visitor due to what was perceived as being squalid living conditions. McClintock, on his fourth Arctic expedition, was more respectful in his descriptions, although admitting that when they employed a young man as their dog-driver he was 'cleansed and cropped ... soap and scissors being novelties to an Esquimaux'.[97] Further north along the west coast of Greenland, McDougall recorded his impression of the appearance of two Inughuit men and three boys near Cape York (Perlernerit): 'they outvied all we had previously seen in want of cleanliness, and were, without exception, the most disgustingly filthy race of human beings it has been my lot to encounter'.[98] Such statements often carry an underlying tone of delight as if their judgement confirms their own superiority.

Terms like 'specimen' and 'race' were common in ethnological discourse at the time. 'Race' referred to human variation, though increasingly during the 1850s and 1860s a 'new tide of Victorian racism' meant that the conceptualisation of the term became associated with fixed, inherited traits that could not be changed.[99] Prichard variously used the terms 'race', 'type', and 'tribe' in his chapter, although he also referred to nomadic peoples as 'savages'; the term 'specimen' was used with reference to collecting human skulls, if it was possible to acquire any.[100] Visual assessment by Goodsir, a surgeon, describing the people encountered at the

Whale Fish Islands (Imerigsoq), tends to take the form of natural history description, to which ethnology was closely related: 'the large head, with narrow retreating forehead, strong coarse black hair, flat nose, and full lips, with almost beardless chin'.[101] This analytical description of the Inuk as a specimen distances the reader, and indeed the author, from any personal engagement with the people. Goodsir homogenised Inuit, separating his ethnological portrait from the narrative, a technique that is noted by Mary Louise Pratt.[102] These descriptions were designed to focus on difference and fed into the hierarchical categories of racist ideology.

The illustrations in the narratives of some visitors to the Arctic mirror this visualisation of the Indigenous person as a type, and depictions of unnamed individuals like that of Inglefield's narrative are common, stripping the person of their individuality. Such images foreshadow the later dehumanising photography of Indigenous peoples, such as that of the 1860s taken in South Africa according to the instructions of Thomas Huxley. Huxley complained that many photographs, which had been taken by travellers, were useless for discerning their physical characteristics and advocated photography of models unclothed with a measuring stick, facing forward and in profile.[103] The female Inuk illustrated in Inglefield's narrative is a servant, unnamed and anonymous.[104] Here, clothing and physical appearance dominate, emphasising features that are characteristic of a group and not of an individual. Her docile appearance is further domesticated by the tray she carries, complete with the European trappings of wine glass and bottle, in place of traditional Inuit artefacts.

For Atlantic-approaching expeditions that overwintered in the Arctic (such as the large Austin and Belcher expeditions), their winter sojourn, commonly in the Parry Channel, was far removed from the possibility of extended interaction with local populations over long periods. This was because the nearest Inuit populations resided to the east, on the coast of Greenland, and on the more southerly islands of the Canadian archipelago. As Cavell notes, the northern half of the Canadian archipelago was abandoned by Inuit in the early part of the Little Ice Age (*c.* 1300–1870);[105] thus the narratives' assessment of the Indigenous people usually ends on the west coast of Greenland, and the empty environments thereafter represented were indeed largely devoid of a local population for much of the year.

Expeditions that approached the Northwest Passage from the west via the Pacific Ocean and Bering Strait had more prolonged contact with Indigenous peoples. The expeditions often wintered near Chukchi and Iñupiat settlements, where prolonged interaction took place over the winter or even successive winters.[106] Such interactions are evident in Hooper's *Ten Months among the Tents of the Tuski*,[107] written by the lieutenant of the *Plover* who was in the Arctic from 1848 to 1852. The bulk of his narrative concentrates on the time the expedition spent with the Chukchi and is clearly reflective of the extensive hospitality he experienced. Hooper's written descriptions of individual Chukchi are character sketches as well as being physically descriptive. Furthermore, the inclusion of *Ahmoleen's Map of Behring's Straits* signals the extent to which expeditions benefited from Indigenous knowledge.[108] The plates show an Arctic environment that includes Chukchi and expedition members, while the value of the social interaction with the Chukchi is evident both in the language used and the illustrations.[109] For Hooper, the environment and people's 'many points of striking interest had become familiar', but a 'new and delightful spectacle' was provided when 'to crown it all, the new arrivals, with their reindeer, filled up the picture which needed nothing more to complete its picturesque and peculiar beauty'.[110] The use of the term 'picturesque' here equates to the unthreatening novelty and pleasing irregularity associated with Uvedale Price's interpretation of the aesthetic.[111]

Although the scene was described as a 'tableau' featuring these new arrivals, this is a scene that includes 'the crew, clad in all varieties of costume, from the semi-military to a close proximity with the dresses of our friends, mingling with the Tuski in their curious habiliments'.[112] These figures in *Winter Quarters, Emma's Harbour* (10.5 × 17.5 cm) are solid and at ease in their environment, showing none of the struggle we associate with some of the expedition figures in other prints. This heavily populated landscape is starkly different to the empty landscapes that Pratt suggests were produced by 'the European improving eye' of late eighteenth-century writers in Africa.[113] It is also distinguished by the inclusion of expedition members as part of the 'tableau', marking it out as different from the *tableaux* of purely Indigenous people that were popular metropolitan exhibitions at the time. Furthermore, the ship, covered in for winter, is shown in the background. Unlike representations of the Arctic in summer, which often show land viewed from the ship's deck, here a view of the ship is seen from land, as the residents would have seen it, albeit from a raised perspective.

This depiction of visitors and residents mingling is also in sharp contrast to an illustration of officers much later in the narrative when a more difficult period in the voyage was experienced – Hooper's stay at New Fort Franklin, when a small party of men wintered in relative isolation in the continental interior. The illustration is small and cramped, the depiction of the officers at one of the log cabins in the interior showing tiny figures, dwarfed by dark fir trees, indicative of the difficult period in Hooper's voyage that this picture represents.[114] Unusually, the book includes an engraving of a drawing done by Enoch, another Chukchi friend, of Hooper himself, providing a rare example of the lens turned on the author (10.5 × 17.5 cm).[115] This is the most intriguing illustration in Hooper's narrative, the portrait of the 'explorer' himself, appearing as helpless and small as an infant and in need of care and assistance, perhaps reflecting the Chukchi view of the outsiders.[116]

Interaction with Indigenous peoples provided relief from the purely homosocial environment of the ship, and Indigenous women were subjected to additional aesthetic and sexual appraisal. The appeal, or otherwise, of their appearance was commented upon, and they are sharply divided into categories based on age. Goodsir, who describes Inuit men in Greenland in scientific, objective terms (his focus on their physical characteristics belies his medical background), shifted his attention when it came to Inuit women, who were rather judged in terms of their visual appeal, or lack thereof: 'A number of Esquimaux women were standing on the rocks when we landed. Some of the oldest of them were certainly the most hideous-looking creatures I ever saw, although one or two half-caste girls amongst them were almost comely.'[117] McDougall, too, commented on the older women unfavourably at Cape York (Perlernerit) in the north of Greenland:

> The inhabitants consisted of two old women, who might have been belles in their younger days; if so, their present personal appearance would tend to prove beyond a doubt that beauty is but fleeting. Three younger and more comely women, each with a child at her back, were presumed to be the wives of the only three men we observed.[118]

This is a common way of looking in the narratives, in that there are two types of gaze cast upon Inuit. The first gaze judges whether the Inuit man can be transformed into a 'civilised' member of society; it compares the Inuit man to oneself. The second gaze is aesthetic or sexual. While young women are often described as 'comely' in the narratives, older Indigenous women tend to be regarded as 'hideous', a term that does not generally attach itself to descriptions of older men. The use of the word comely, a common nineteenth-century term indicating a person of pleasing and pretty appearance, describes a more picturesque appearance that is quite different to beauty, but nonetheless pleasant to look at, much in the same way that an irregular landscape might be picturesque but not beautiful. In the nineteenth century, the term 'comely' was applied to men as well as women.[119]

Hooper veered close to the supernatural in his description of an old woman when he described a traditional dance where 'an aged woman with shrivelled limbs, and hideous puckered visage, essayed a feeble exhibition, crooning out also in a thin and shaking tone, and concluding her deed of might with a grin of horror'.[120] This 'exhibition' inspired a range of emotions, including pity, although Hooper's old woman was ridiculed in the end: 'we laughed loudly and long, at which . . . [she] cackled to herself in high glee'.[121] The description of the cackling old woman – 'hideous', with her 'grin of horror' – aligns itself to the witch archetype of fairy tales.

Although Hooper continued to judge women by appearances, he also considered the character of individual Chukchi women. He considered that his 'friend' Yaneenga was 'one of the best and worthiest specimens of her tribe', although other, younger women might 'dispute with her the palm of beauty . . . but, who, like Yaneenga, bore so unvaried a countenance of good-humour? – who, like her, was always amiable, always thoughtful for the wants or comforts of those around her?'[122] While Hooper referred to her as a 'specimen' of her 'tribe', using the language of ethnology, he praised her character above appearance and thought of her as a 'faithful friend'.[123] She and her family appear throughout the relevant portion of the narrative as important and respected figures. These connections with the Chukchi were held in high regard by Hooper, who quotes an early entry from his 'private note-book' in his narrative, to show how his impressions of the Chukchi were rooted in what he referred to as friendship: 'Really we are becoming quite domesticated with these people; they visit our mess-room, and go from cabin to cabin, eat with us, drink with us, and are exceedingly good friends.'[124] The use of the term 'domesticated' suggests Hooper's association of the Chukchi with wildness and savagery, and yet his descriptions of Yaneenga hint at the possibility of something rare in the contact zone: a 'genuine interpersonal relationship'.[125]

The character of Yaneenga and others in the Bering Strait area make it all the more noticeable that the lower orders on the ship itself do not merit as much attention in the narratives. The seamen on the ships were the lowest rank on a vessel and appear only as shadowy figures in the background of texts. This group, spending far more time with other ranks during an Arctic winter than on a normal voyage, are an often juvenilised generic mass. We might even think of winter quarters as another type of contact zone, where officers and seamen came into deeper contact with each other. Elusive to the modern reader, they are known to us only through general comments. Osborn, for example, juvenilised the sailors when he described a lieutenant relating the voyages of Parry to them during the winter of 1850 to 1851: 'In an adjoining place, an observer might notice a tier of attentive, upturned faces, listening, like children to some nursery-tale.'[126] Indeed, certain Indigenous individuals, in some cases, can be treated with more respect than seamen on the expedition, for their ability to survive in the Arctic as well as for personal characteristics. Seamen, on the other hand, so integral to the expedition, are sometimes portrayed as children, finding fun when they can by sliding down icebergs and praised when they labour in the sledge harness.

An incident related by Goodsir early in his narrative reminds the modern reader of the status of sailors and class divisions on board when two unnamed sailors are lost overboard during an Atlantic storm. The event, while regrettable, does not warrant more than a few lines, and Goodsir purported that the other men soon forgot the incident: 'sailor-like, in a few days all was forgotten'.[127] In common with many other authors of Arctic travel narratives, the sailors here are referred to as 'Jack', it not being necessary to name them individually.[128] Their individual invisibility suggests that the common seaman was sometimes regarded as an expendable item. In Osborn's narrative, we see none of the 'red-faced mortals, grinning, . . . the men labouring and laughing; – a wilder or more spirit-stirring scene cannot be imagined'.[129] Neither is the 'bustle and merriment' or 'gambols of divers dogs . . . with small sledges attached to them . . . amidst shouts of laughter' shown in the plate *Sledge Travelling*.[130] Pictorially, the seamen are only represented as a labouring group, their individual faces invisible to the reader. These men have become representatives of the difficult labour of searching for Franklin.

While first encounters with Indigenous populations were often made visible using descriptors like 'filth', they were also viewed in a more scientific way, aligning to the tradition of natural history and the new discipline of ethnology. Indigenous women were made visible in terms of age and appearance; however, long-term contact led to increased interaction in the contact zone. The *Plover*, for example, was stationed in the Bering Strait region, with some expedition members spending six years in the Arctic. In contrast, the common sailor of the expeditions is almost invisible in the Arctic narratives of exploration, appearing as a group rather than as individuals in their own right, suggesting that some Indigenous individuals were held in higher regard than British seamen.

Intertextuality, Transformation, and Pictorial Commodification of the Landscape

Representations of landscapes change as they move through space and time; the depictions of the Arctic are not fixed, but continuously fluctuating. Sketches were remediated by engravers, lithographers, and publishers. Sometimes, illustrations were created from the author's textual description, and the original sketch was not necessarily made during the expedition.[131] As Qureshi notes (with respect to representations of people), picture and context could easily become divorced from each other in the nineteenth-century circulation of pictures; furthermore, one of an illustrator's priorities was 'the creation of a memorable image employing symbols commonly associated with their subject, even if it was at the expense of obvious factual accuracy'.[132] The same holds true for the depiction of Arctic environments in print, with the symbols of polar bears, ice, shipwreck (or possible shipwreck), and man-hauled sledge travel occurring repeatedly. As Keighren, Withers, and Bell show, there was a sense among exploration authors in the 1850s that the reading public was 'saturated with sublime imagery', 'indifferent to faithful representation', and that 'lithographers had become accustomed to exaggeration'.[133] Publishers, for their part, were keen to make a profit and may have directed lithographers and engravers to heighten the impact of a picture.[134]

The development of new printing techniques and the increased production of illustrated books and periodicals in the nineteenth century is central to contextualising the production of Arctic travel narratives. In keeping with natural history books published in the period, Arctic narratives utilised tinted lithographs, colour lithographs, woodcuts, wood engravings, and mezzotints. Several printing methods are sometimes used in the one book, for example in Kane's first Arctic narrative, which contained mezzotints, lithographs, and wood engravings.[135] Mezzotints facilitated a wide tonal range, making the rendering of contrasts subtle and atmospheric. When comparing publications that are a decade apart, we need to be mindful of such matters as printing costs, the number of colours used, and factors governing the production of the prints. For example, while it was common for professional artists like J. M. W. Turner to have their paintings converted into prints, they would have personally overseen this transformation. A naval lieutenant and amateur artist did not have the same power over his artwork once the expedition returned, particularly when he was not the book's author, as was often the case. Thus, prints, in their varying forms, do not necessarily represent expedition members' visual perceptions of the Arctic but were mediated by engravers, lithographers, and publishers.

In Osborn's *Stray Leaves from an Arctic Journal* (1852), which went through several editions, the first edition had a folding map, four colour lithographs, and five monochrome engravings.[136] The inclusion of a map was seen as a 'warrant of authenticity and of accomplishment' by the explorer.[137] In the preface, Osborn acknowledged the work of two other officers who were on the voyage, thereby also conferring authenticity on the illustrations: 'To Lieutenant W. May and Mr. McDougall, I am much indebted for their faithful sketches.'[138] The illustrations were printed by the well-known lithographic printing firm, M & N Hanhart, who specialised in natural history plates, landscape scenery, and pictorial covers of sheet music for popular songs in the Victorian period.[139] Should the Victorian reader only choose to look at the plates in this book, they would already gain a good understanding of some of the themes of a maritime Arctic voyage. Many of the crucial signifiers are here: exploration, as represented by landmarks; possibility of shipwreck, depicted by ships breaking through sea-ice; and masculinity, represented by sledge-hauling in a snowstorm.

The frontispiece, *Cape Hotham* (9.3 × 15 cm), depicts a topographical landmark on Cornwallis Island, Barrow Strait, deep in the Canadian Arctic. The short caption is likely aimed at the 'general reader'[140] to whom Osborn referred in his preface. At the same time, naming the location lends it authenticity. Although the scene is relatively tranquil, there is a significant quantity of sea-ice evident. The sky is tinged with evening hues of yellow and pink while the black smoke, connecting to the overall narrative that emphasises the use of the screw steamer, is clearly seen rising from the tender's funnel. It is likely this picture represents an occasion in August or September 1850 when the ships were in the area.[141]

There are several points of intertextuality when we examine the subject matter (places and events) of the plates. This frontispiece, *Cape Hotham*, has a parallel representation in the *Illustrated Arctic News*, one of the newspapers of the same expedition, in which the picture *U.S. Brigs – Advance & Rescue, Passing Cape Hotham*[142] (6.7 × 13.5 cm) illustrates the ships of the American expedition financed by Henry Grinnell, a wealthy American businessman who funded several private Arctic expeditions. The caption on the latter picture is more specific and refers to events within the environment, not only a notable landmark. The presence of the American expedition is also referred to in Osborn's narrative: 'Commander De Haven's gallant vessels, who, under a press of canvas, were just hauling round Cape Hotham'.[143] The two pictures are strikingly similar and represent events that took place within days in September of 1850. However, some obvious differences are evident. The illustration from Osborn's narrative shows a greater quantity of sea-ice. Additionally, the steam tender and its smoke catch the viewer's eye in the narrative. Cape Hotham itself is shown from a slightly different angle, an angle that emphasises its distinctiveness as a landmark and makes it appear higher. The transformations are fairly subtle, however, and a link to scientific recording and a commitment to topographical accuracy show the strong tradition of naval surveying in Britain.

Reiterations of other landmarks such as the Devil's Thumb and what would become well-known sights like the three graves at Franklin's first winter quarters, in various narratives and in other illustrative formats, give us a sense of the fluctuating visualities of the Arctic during this period. For example, the graves at Beechey Island, the first location of a clue to the mystery of the Franklin disappearance, became a key site in the story of the Franklin searches.[144] The graves with their wooden headboards were first discovered in 1850 and during that summer became a macabre tourist spot with no less than ten manned search ships, including over three hundred expedition members, in the area.[145] This site was the location of the winter quarters of Franklin's expedition in the winter of 1845 to 1846 and, as well as the graves of three of Franklin's men, presented traces of a workshop and garden.[146] Whether depicted dramatically, playing to the darker aspect of the Franklin search, or sketched as a naval record, the pictures of this site tell us much about the processes of transformation underway in Arctic imagery at this time.

Starkly contrasting representations of this scene exist. Its first public appearance was in the *Illustrated Arctic News* where the monochrome print, *Seamens Graves – Beechey Island* focuses on the graves themselves and shows merely the slopes of Beechey Island in the background.[147] It was originally hand-drawn by McDougall on the Austin expedition and was then entitled *The 3 Graves – Beechey Is.d. The Winter Quarters 1845–6 of the Expedition under Sir John Franklin* (8.8 × 14.8 cm);[148] the pen-and-ink drawing shows his careful, characteristically neat, and finely delineated line, suggesting attention to detail, with the names of each man written below his respective grave on the picture.[149] The picture is more a technical drawing and has no aesthetic aspirations, being a visual record of the graves rather than an emotive response.

The site took on a far more sinister and gothic appearance when it appeared in Kane's bestselling romantic narrative that made him the 'icon'

Figure 3.4 James Hamilton, after Elisha Kent Kane, *Beechey Island – Franklin's First Winter Quarters*, 1854. Mezzotint, 10.4 × 16.2 cm. Plate 4 of Kane, *The U.S. Grinnell Expedition in Search of Sir John Franklin. A Personal Narrative* (New York: Harper & Brothers, 1854). Courtesy of Smithsonian Libraries and Archives / Biodiversity Heritage Library. Public Domain.

of the search.[150] Over several years, Kane had been encouraged by his family to write up his escapades as a young doctor travelling the world, and by the time he went to the Arctic in 1850, he had become adept at turning any story into a thrilling adventure with a veneer of scientific authenticity.[151] This plate (Figure 3.4) was one of those engraved by the notable mezzotint engraver John Sartain from drawings by James Hamilton. The subtitle is careful to include the words 'after a sketch by Dr E K Kane', conferring authority on the depiction. Entitled *Beechey Island–Franklin's First Winter Quarters*, the caption fails to refer to the graves. However, the silhouetted grave markers are dramatically backlit and clearly are the focus of the picture, lending a ghostly atmosphere to the scene. The half-concealed moon adds a sense of mystery, and the peaks of an iceberg on the horizon recall gothic edifices.

The castellated iceberg is in fact quite out of place here; Beechey Island is located on the south-west tip of Devon Island, and the glaciers are all to the east. The icebergs they created were swept out eastwards by the current to Baffin Bay. Any icebergs that enter from Baffin Bay only move as far westward as approximately 85°W, whereas Beechey Island is located at

91.50°W.[152] The slopes of Beechey Island are here represented as vertical cliffs, implying that the nature of the landscape is dark and unfriendly. Most curious of all is that the picture shows a night-time scene, although the location was visited in the bright Arctic summer. The choice to re-imagine the scene in darkness was a calculated one, prioritising sensationalism over science, presenting a far more romanticised, gothic, and obscure Arctic. At the same time, the rather haunting beauty of the scene is aesthetically appealing.

Although the graves are not mentioned in the caption, the eye is drawn to their dark silhouettes, adding a sense of mystery and even antiquity to the piece. The individuals buried do not seem important, unlike in McDougall's sketch where they are named; rather it is the ghostly atmosphere created by graves that lures the viewer in. Although Kane included the grave inscriptions in the narrative, our first introduction to them is when a messenger comes to Captain Penny, who is with Kane on Beechey Island: 'The news he brought was thrilling. "Graves, Captain Penny! graves! Franklin's winter quarters!"'[153] The lines coalesce with Lambert's assessment of the narratives of the searches as retelling a 'big boy's adventure'.[154]

The Devil's Thumb is a feature that appears represented differently in several narratives. Seen twice in Kane's narrative, the Devil's Thumb here is tall and sharp, dark and angular, and placed in the centre of the composition.[155] In William Henry Browne's lithograph *The Devil's Thumb, Ships Boring and Warping in the Pack* (12 × 18.5 cm),[156] the landmark is far in the background behind the main activity of the ships in the ice-pack; even accounting for perspective, it is far less dramatic and vertical than in Kane's illustration. The caption on Browne's version includes events within the Arctic environment, whereas Kane's brief caption only references the landmark's unusual name, emphasising its novelty. A prose description of the Devil's Thumb also appeared in Jules Verne's *Captain Hatteras*, the fictional account of an imagined Arctic voyage, first published in French in 1864. Referencing the voyages of both Kane and Snow (and by implication their narratives), Verne used the landmark to set a bleak, yet supernatural, stage: 'The odd form of the Devil's Thumb, the dreary deserts in its vicinity, the vast circus of icebergs . . . rendered the position of the *Forward* horribly dreary.' Near the 'ever-threatening Devil's Thumb', the ship is put in the 'critical position' so familiar to the readers of the Arctic exploration narratives. Here, the icebergs 'drifted by in the fog like phantoms'. 'Sometimes, under the action of the storm, the fog was torn asunder, and displayed towards the land, raised up like a spectre, the

Devil's Thumb.' The chapter culminates with the reappearance of the captain's menacing and seemingly supernatural dog, which some members of the crew had attempted to murder. This gave the 'finishing touch to their mental faculties'.[157] Verne's Arctic narrative shows an acute awareness of the atmosphere evoked by the illustrations in Kane's narratives, the ghostly, other-worldly space that challenged men psychologically as well as physically.

A more dramatic Arctic began to creep into the narratives as time went on, partially influenced by the success of Kane's narratives but also perhaps by Arctic panoramas that were exhibited during the 1850s. McClintock's *Voyage of the 'Fox'* (1859), which went through several editions, included illustrations by Walter May, who did not participate in the expedition. A third edition of *Voyage of the 'Fox'* (1869) contained additional illustrations, many of which are of a more dramatic nature than in the first edition. In this later edition, we find that one of the earlier plates has also been removed.[158] Walter May's *Moonlight in the Arctic Regions*,[159] which showed a view of a tabular iceberg[160] and a ship in the first edition of McClintock's narrative, was replaced by an unattributed illustration captioned *Moonlight – August 25 1857* in the later edition.[161] Although the date within the caption now attempts to lend authority to the picture, the Arctic has been utterly transformed. The new, heightened contrasts of light and dark are unlikely to have been so extreme at high latitudes in August, and the more ordinary, horizontal icebergs have been replaced by a dramatic pinnacle iceberg, which appears as a vertical and gothic fortress of ice, one that bears a striking resemblance to the mass depicted in *Melville Bay* in Kane's *The U.S. Grinnell Expedition* (1854) and is not dissimilar to the gothic iceberg in *Beechey Island*.[162] The clear full moon of *Moonlight in the Arctic Regions* has become a semi-concealed body, adding an atmosphere of mystery to the Arctic and increasing the resemblance to the moon depicted in Kane's *Melville Bay*.

This transformed environment, where dramatically lit Gothic fortresses hint at mystery, suggests new wonders beyond the horizon, and the vertiginous treatment of the ice is key to this gothicised Arctic. Kane's illustrations are often untitled vignettes that lend the narrative a dream-like quality and place emphasis on the narrative as adventure story, aligning more with Verne's *Captain Hatteras* than with an exploration narrative. The illustrations of Kane's narratives were carefully planned and executed to maximise the sensational elements of the Arctic expedition. So, while British narratives took some liberty in depictions, including alterations from drawing to print, Kane's narratives reconstructed the Arctic space completely, moving land masses to where they might be better suited (for example, shifting the sea cliffs found in Lancaster Sound closer to the graves), exaggerating heights, and adjusting the contrasts of light and dark to suit a more gothic mood, blurring the boundaries between fiction and 'reality'.

Conclusion

McClintock's mind was 'busy with some sort of magic lantern representation of the past, the present, and the future'[163] as he attempted to sleep during the frustrating summer of 1856, when his ship, the *Fox*, was delayed by pack-ice. The visual culture of the mid-nineteenth century permeated the writing of even the most pragmatic authors of Arctic exploration and travel narratives, affecting not only the ways in which they told their stories for readers but also their own experience of the Arctic. Other authors went to greater lengths, visually inviting readers to experience panoramas, exhibitions, and *tableaux* through their descriptive passages, as they drew on Romantic poetry and the power of memory to create an Arctic experience for readers.

The Arctic exploration and travel narrative blended realism or factual writing with elements of gothic fiction and more touristic travel writing. Authors crafted writing that emphasised danger, masculinity, and discovery, with disastrous expeditions tending towards the most extreme depictions of the Arctic, positioning the environment as a formidable enemy. Winter darkness, improbably formed icebergs, and a sense of the unknown, sometimes heightened by deprivation, led some narrators to invite their reader into an unfamiliar and unstable world that bordered on fantasy. For some writers and artists, the Arctic had positive fairy-tale-like qualities, with icebergs fit for a 'fairy-queen';[164] for others the ice was 'rude and grotesque'.[165] Perhaps unsurprisingly, the most polarised reactions to ice and the Arctic landscape in published narratives were from first-time visitors or from those who only spent a summer in the Arctic. In many cases, the Arctic environment was responded to aesthetically and subjectively. Authors on ill-planned disastrous voyages also described the ice in exceptional aesthetic terms, lending the apprehension of the negative sublime to their narrative and producing the most vivid and self-consciously literary picturing of the land, sea, and icescape.

However, for those maritime expeditions that wintered over repeatedly with some success, the ship frozen into the ice afforded stability and became a 'home'. The surrounding environment became a locality with feelings and memory attached. Pratt outlines the 'standard elements of the imperial trope' as 'the mastery of the landscape', the 'aestheticizing adjectives', and the 'broad panorama anchored in the seer'.[166] But for authors like Osborn and Hooper, winter quarters became something like a 'home', albeit one where they performed the role of local guide for the reader. For these authors, the panorama was not anchored in them, but they became anchored in the panorama, involved in and connected with the natural environment. When ships wintered near Indigenous communities, extensive interaction could take place, and their presence is vividly represented in texts like *Tents of the Tuski*, in contrast to the seamen aboard ship who remained a faceless collective in the narratives. Expeditions benefited enormously from Indigenous Arctic residents, while not openly acknowledging that, and good relations were fostered to ensure trade, knowledge acquisition, skilled labour, social interaction, and in some cases, sexual relations.

The Arctic, a region that few nineteenth-century readers could ever hope to visit themselves, lent itself well to restructured topography, exaggeration, and mobility, particularly following the success of Kane's narratives, which successfully gothicised the region and encouraged other publishers to follow suit.

CHAPTER 4

'Never to Be Forgotten'

Presenting the Arctic Panorama (1850)

When Jane Franklin arrived to inspect Robert Burford's newly painted *Summer and Winter Views of the Polar Regions* in February 1850, she was joining the ranks of thousands of people who attended such visual events. As the nineteenth century progressed, visual culture became easier and cheaper to access, and spectacles such as panoramas sought to create an experience that offered more than simply viewing an engraving or a framed painting. The panorama placed the spectator at the centre of a circle, surrounded by a huge painting on a curved surface.[1] These large 360-degree paintings became an international phenomenon, and many other visual spectacles evolved out of the original late eighteenth-century concept.[2] By the mid-nineteenth century, the standard entry fee into a panorama had fallen to one shilling, with half-price entry for schools and children, making it an accessible experience across a broad spectrum of society.[3]

This chapter focuses on the large, and very popular, Arctic panorama *Summer and Winter Views of the Polar Regions* (1850), which opened in London shortly after the return of the first naval search expedition for Franklin, the voyage of 1848 to 1849 led by James Clark Ross. The panorama was exhibited at the Leicester Square rotunda, then run by Robert Burford, in February 1850. This respected establishment first introduced the concept of panoramas to London in 1793 and was seen as a venue of 'instruction as well as entertainment', which depicted places 'so accurately that you may well suppose yourself transported thither'.[4] By making extensive use of contemporary reviews, prints and sketches, and both published and unpublished written sources, I offer in-depth analysis of the content and reception of *Summer and Winter Views*.[5] This panorama was based on the sketches of William Henry Browne, late lieutenant of the Ross expedition, and its reliable source was advertised repeatedly, with Browne even writing a letter (from an address in Liverpool) to *The Times* on the matter, thereby signalling its authenticity in comparison to 'several exhibitions' that were 'purporting to show views of the Arctic and Polar Regions'.[6] A concern with 'truth' and 'accuracy' pervades contemporary reviews of visual exhibitions like panoramas, and those that announced their origins from on-the-spot sketches reassured viewers that their contents were not deceptive. Reviewers' preoccupation with the 'truth' and 'reality' confirms the power of Burford's panorama to be regarded as a source of information, yet also belies a distrust of popular visual culture in general. Contemporary reviewers praised *Summer and Winter Views* for what they considered to be its realism and truthfulness, but simultaneously noted its 'terrible and fantastic' icebergs and its 'supernatural aspects', implying that such a place could not be real.[7] However, as I will show, the Arctic represented in the panorama, thought to be 'so very real',[8] was drastically different to that represented by Browne and to that conveyed in travel narratives and unpublished written records. Ultimately, *Summer and Winter Views* transported its viewers to a more sensationalistic and supernatural space that downplayed geography, domesticity, and reality, in favour of gothicised icescapes, masculine endeavour, and meteorological effects that combined to create a theatre for the moral sublime.

Summer and Winter Views formed part of a wider visual response to the early stage of the search expeditions, one that emerged through some of the many 'public amusements' produced for the Victorian urban-dweller.[9] In London alone, a growing choice of large Arctic representations was simultaneously available to the public. In November 1849, a new play, *The Sea Lion; or, the Frozen Ships and the Hermit of the Icebound Bay*, had gone to 'considerable expense in getting up the scenery'. This depicted several Arctic views, giving 'a forcible idea of the horrors the brave explorers [on the Ross expedition] must have had to encounter'.[10] By December, *View of the Polar Regions* was part of the new attractions at the Colosseum, Regent's Park.[11] At Minerva Hall, the *Grand Moving Panorama of the Arctic Regions* was among the Christmas exhibitions; it 'combined the results of the principal Arctic navigators', and its 'faithful scenery' was 'authentically rendered from original drawings'.[12] *Summer and Winter Views* opened in February 1850, and in March a series of dissolving views of the Arctic Regions 'with an interesting description' was exhibiting twice daily at the Royal Polytechnic Institution.[13]

Until recently, this panorama, known mainly through an engraving in the descriptive booklet available at the exhibition (Figure 4.1), had not received any in-depth critical attention. Laurie Garrison notes that in the 'current historical literature, any individual panorama usually only receives a few sentences or a few paragraphs of attention'.[14] She addresses this by

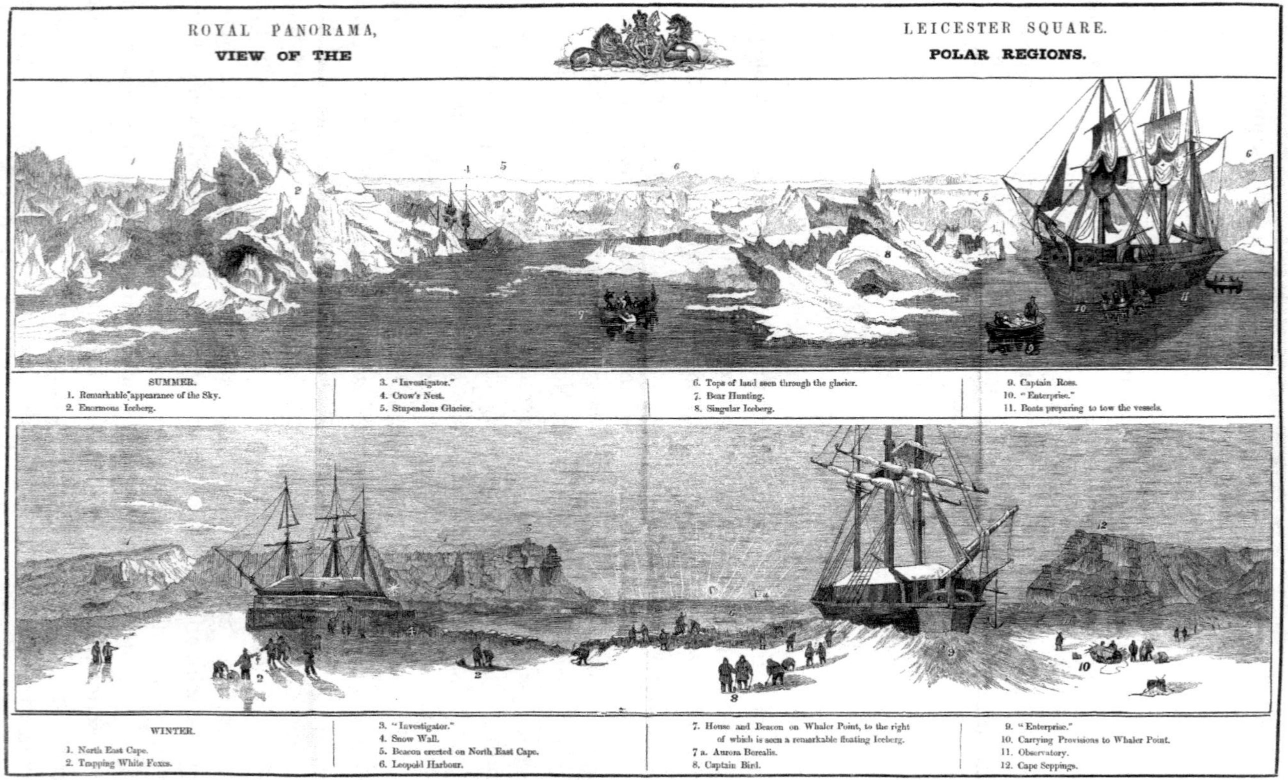

Figure 4.1 Page from Burford's booklet (1850), showing the visual layout and key to *Summer and Winter Views of the Polar Regions*. Engraving, 26 × 42.7 cm. Courtesy of Toronto Public Library.

examining the 'scientific and political context of the production, exhibition and reception' of *Summer and Winter Views*.[15] Garrison maintains that the Admiralty and the proprietor of the panorama collaborated to portray a version of the lately returned Arctic expedition that exonerated those involved, 'having been unavoidably thwarted by the powers of nature'.[16] This was, she argues, evident both in the content of the panorama, which emphasised the activities of expedition members, and in the text of the accompanying booklet that offered a 'firm justification' of the Ross expedition and the Admiralty.[17] However, she notes that the panorama encouraged multiple responses, especially an emotional response, as evident by reviewers' reactions, concluding that 'multiple interpretations were at work, undermining any singular, hegemonic view'.[18]

Russell Potter briefly discusses *Summer and Winter Views* in the broader context of nineteenth-century Arctic spectacles, noting how Burford had 'divided time and space in two'[19] and surmising that this reflected the 'curious doubleness of the northern sublime – a sublimity of chaotic action, lethal in one moment and picturesque the next'.[20] Robert G. David deals with the panorama in his general survey of nineteenth-century British visual representation of the Arctic, arguing that Burford intentionally depicted both the sublime and the picturesque by dividing the Arctic into two halves.[21]

This chapter builds on the work of previous critics, offering a more art historical approach to the content of *Summer and Winter Views*, attending to primary visual sources and a wider range of written sources, noting in particular how certain aspects of the experience of expedition members were erased, replaced, or transformed to create a more supernatural, gothic, and masculine space.

The Panorama *Summer and Winter Views of the Polar Regions* (1850)

In 1793, Irishman Robert Barker opened a permanent exhibition centre for his panoramas in the form of a rotunda in Leicester Square, London, which continued in business for seventy years. He was succeeded by his son, Henry Aston Barker, in 1806 and eventually by Robert Burford, who had worked with them, in 1827.[22] By the mid-nineteenth century, London was awash with visual experiences, so much so that Burford felt compelled to state 'with the exception of one, it is the ONLY PANORAMA IN LONDON, though various other exhibitions, consisting merely of moving pictures, make use of the term Panorama'.[23]

In December 1849, the *Athenaeum* commented: 'There is a perfect battle of *B*'s with new panoramas to catch the holiday people of London and the country visitors of this festive period of the year. We have Burford, Banvard, Brees and Bonomi, all catering successfully for the amusement and instruction of the public.'[24] Such exhibitions were expected to be educational as well as entertaining, making their association with reality an important selling point. As Sadiah Qureshi points out, 'pleasure and instruction were neither mutually exclusive nor necessarily confined to distinctly separate spaces'.[25] The panoramas were thought of as 'Animated Illustrations of Geography' and were considered to be a 'vehicle for instruction' as well as 'an illusion to the senses and a new luxury in aesthetical art'.[26]

The proprietors of the Leicester Square rotunda tended to emphasise the quality and accuracy of its painted images, as well as their educational value, over the showier imitators. They traditionally painted their panoramas based on their own on-the-spot drawings of places they had travelled to, while similar spectacles tended to employ dramatic lighting, moving props, and elaborate illusory effects in order to draw the crowds.[27] Robert Burford and his assistant Henry Selous both exhibited paintings at the Royal Academy,[28] and Burford felt it his 'duty' to emphasise that his panoramic views were not 'a species of scene-painting, coloured in distemper, or other inferior manner' but were 'painted in the finest oil-colour and varnish ... and in the same manner as a gallery picture'.[29] Despite the competition, the original panorama, run by Burford from 1827 to 1863, was more valued by some. The art critic and theorist John Ruskin later expressed his feelings on the varying quality of panoramas: 'Calame and that man – I forget his name – are merely vulgar and stupid panorama painters. The real old Burford's work was worth a million of them.'[30] He lamented the closure of Burford's panorama in 1863, suggesting that it 'ought to have been supported by the Government'.[31] The closure was equally regretted by the writer of *Leicester Square; Its Associations and Its Worthies* (1875) who felt that it was 'a real loss, ill-supplied by the ever-increasing swarm of picture exhibitions'.[32] Around the time of the exhibition of *Summer and Winter Views* a writer for *Musical World* referred to Burford's panoramas as 'decidedly the highest works of their class, – where we contemplate Pompeii, the Arctic Regions, and the Lakes of Killarney'.[33] The subject matter treated by the panorama encompassed scenes of battle, voyages of discovery, views of cities, localities of natural beauty, and public ceremonies.[34] It was not only a 'means of virtual travel' but was also a 'vector of news', showing natural events and battles that celebrated 'a sense of British military might and identity'.[35]

Burford's panorama in Leicester Square was far more than an amusing spectacle. Ruskin considered it to be 'an educational institution of the highest and purest value . . . one of the most beneficial school instruments in London'.[36] Alexander von Humboldt believed that panoramas should be used to depict nature, and that if large panoramic buildings containing 'a succession of such landscapes, belonging to different geographical latitudes and different zones of elevation, were erected in our cities and . . . thrown freely open to the people, it would be a powerful means of rendering the sublime grandeur of creation more widely known and felt'.[37] Humboldt singled out Barker's panoramas when talking about the improvement in landscape painting on a large scale, as being a 'kind of substitute for wanderings in various climates'.[38] A plate from an early nineteenth-century book[39] shows a section of the Leicester Square rotunda with its panoramas and visitors, the majority of whom are women (often being guided by men), a reflection perhaps of their more geographically constrained role in nineteenth-century society.[40]

By the time *Summer and Winter Views* was exhibited, the rotunda at Leicester Square was three storeys high; the lowest and largest storey, which housed *Summer and Winter Views*, was approximately twenty-seven metres in diameter. The circular balustraded platform from which the polar panorama was viewed was approximately nine metres in diameter and the painting was always viewed from a distance of no less than nine metres.[41] The layout was designed so that customers entered by a series of darkened stairs, emerging on a central viewing platform into an all-encompassing experience, where they had no frames or reference points to distract them from the illusion. The platform on which the spectators stood and observed the Arctic representation must have given them the feeling of standing on the deck of the ship, or even up in the crow's nest; the visual impact of Burford and Selous's representation of the Arctic should not be under-estimated. Viewing their panorama provided an experience for the Victorian senses far beyond that evoked by reading a newspaper or narrative, attending a lecture or viewing a single framed painting in a gallery. The panorama eliminated the frame of a picture that people were used to seeing, as well as removing any reference points of size and distance outside the painting,[42] leading Humboldt to describe the spectator of the fixed panorama as 'enclosed as in a magic circle and withdrawn from all disturbing realities'.[43]

Although it is not known how many people visited *Summer and Winter Views*, such exhibitions were enormously popular in London at the time. On 26 December 1849, 19,986 people visited the British Museum, and the Colosseum, where the Dansons' *View of the Polar Regions* was then exhibiting, received 'an immense number of visitors on the same day'.[44] Madame Tussaud's was so 'densely crowded from morning to night' on the same day that it was necessary to have 'additional police-constables on duty in the street to prevent accidents'.[45] Given that *Summer and Winter Views* received numerous favourable reviews and Burford's establishment was highly regarded and widely known, it is likely that very large numbers of people visited during the fourteen months for which it ran.[46] The private viewing of *Summer and Winter Views*, which Jane Franklin attended, on 9 February 1850 (prior to the public opening) was crowded and included the Lords of the Admiralty among the earliest guests.[47] Two months after the opening, the *Critic* noted that it continued 'to attract delighted crowds'.[48] Nine months after it had opened, it was still making an impression, with the *Critic* declaring that it was a scene 'never to be forgotten'.[49]

Reviewers noted the panorama's artistic qualities, but even more important seems to have been the assumption of its truth value for some spectators, or how closely it mimicked the 'reality' of the Arctic, a place that viewers would never be able to go to in order to verify the representation. Although Burford and Selous often drew from life themselves, by obtaining Browne's sketches from the Admiralty, they had secured the next best thing, a first-hand eyewitness account.[50] The display of furs on the viewing platform, as well as drawings by Browne, was a calculated move to bolster the panorama's authenticity.[51] The numbered key under the engraving of the panorama in the printed booklet (see Figure 4.1) also implied that the panorama was a factual record. But beyond ordinary pictures, Humboldt had noticed the immersive quality of the panorama that caused 'impressions' to mingle with 'remembrances of natural scenes'.[52] The all-encompassing panoramas became part of spectators' memories, both of people who would never go to the Arctic and of future expedition members, like William Parker Snow, who, when he did get to the Arctic six months later, 'almost, fancied that [he] was again in London viewing the artistic sketch'.[53] However, the power of the panorama to immerse the viewer in 'reality' also caused an anxiety around its purpose of deceiving the spectator. In 1849, Professor Charles Robert Leslie gave a lecture on painting at the Royal Academy where he asked 'whether others have not felt what has always occurred to me in looking at a Panorama, that exactly in the degree in which the eye is deceived, the stillness of the figures and the silence of the place produces a strange and somewhat unpleasant effect'.[54] Leslie felt that there was always something '*unpleasant*—in all Art, of every kind, of which deception is an object. We do not like to be cheated.'[55] Indeed, as Nigel Leask observes, it could be thought of as a 'sort of troubling early nineteenth-century version of "virtual reality"'.[56]

Some of the unreservedly positive reviewers of *Summer and Winter Views* were particularly drawn to what they considered to be its 'truthful' qualities, perhaps unconsciously trying to offset this unpleasantness that Leslie identified. The *Athenaeum* considered *Summer and Winter Views* to be 'one of the most successful' panoramas that they had seen.[57] The *Art Journal* declared, 'we have never witnessed more interesting pictures than the present, or any possessing greater novelties in effect . . . Altogether, we do not remember a more peculiarly truthful and artistically beautiful production exhibited by the talented proprietor.'[58] The use of the term 'witnessed' here implies the reliability of an eyewitness account, while the phrase 'peculiarly truthful' suggests the uneasiness associated with deception. The *Critic* noted the difficulty of persuading oneself that it was not, in fact, reality: 'It is so *real* that an effort is needed to abstract the mind from the scene in the picture to the reality.'[59] As late as November 1850, the same periodical commented that Burford's polar panorama was 'a sublime scene, never to be forgotten, it is so very real'.[60]

A large part of the reviewers' conviction of the panorama's realism came from the fact that *Summer and Winter Views* was directly linked with scenes recorded by Browne during the Ross expedition of 1848 to 1849, a connection that was reiterated in reviews and advertisements. The *Era* reasoned that as Browne's on-the-spot drawings had been 'transferred by Mr. Burford to canvass [*sic*], the faithfulness of the scenery may therefore be depended on'.[61] The *Morning Chronicle* noted that the panorama 'professes to be a correct view' and that the 'effects' were produced 'without exaggeration or meretricious display'.[62] The *Observer* reviewer went further, marvelling that, although Burford 'had to labour under the disadvantage of pourtraying [*sic*] scenes which he himself had . . . never seen', they were 'assured by gentlemen who had wintered in these northern latitudes, that the resemblance of the panorama to the reality is perfect'.[63] Unfortunately, we only have access to three rough sketches or paintings by Browne that reside in public archives from the Ross expedition. The 'drawings' displayed at the panorama, whose whereabouts, if they survive, are unknown, were likely to have been more finished versions of sketches done in the Arctic.

Considering that Browne's visual art was the first body of work to be publicly produced from the Franklin search expeditions, remarkably little is known about his life compared to some of his contemporaries. For example, he fails to appear in the *Dictionary of Irish Biography*[64] or in the *Oxford Dictionary of National Biography*.[65] In view of the fact that his sketches inspired a panorama, likely to have been seen by tens of thousands, it is worth taking a closer look at his background. Although several sources record that Browne was from Dublin, William Henry James Browne was actually born in the village of Howth, County Dublin, *c.* 1823, where his father James Browne was the harbour master from 1818 to 1836.[66] At the time, Howth was far from picturesque: in 1825, an anonymous letter to the *Freeman's Journal* described a series of 'disgusting' views in the area, concluding that Howth was a 'totally neglected village' despite the large sums of money spent in the construction of its new harbour.[67] Browne initially joined the merchant navy, but during the 1840s he served as midshipman on the Royal Navy surveying vessels, the *Sulphur* and the *Samarang*, both under the command of Captain Edward Belcher in the Pacific.[68] His watercolours of coastal elevations from this period show great attention to detail, a delicacy of brushwork, and an interest in the work far beyond that required for the task.[69] This earnest dedication to his visual practice is evident particularly in the carefully painted titles on these elevations, an attention to detail that was unnecessary, and unusual, in the production of navigational drawings.

In 1848, Browne was appointed as third lieutenant on the *Enterprise*, on the naval expedition led by James Clark Ross to search for Franklin in the region of the magnetic north pole. The eighteen-month-long expedition spent a winter in the ice pack at Port Leopold, Somerset Island, and the men suffered badly from inadequate provisions. Despite this, Browne must have been sketching regularly as, on his return to England, his visual records were transformed into a folio of lithographs, *Ten Coloured Views*[70] (discussed further in Chapter 5), as well as *Summer and Winter Views*. Although Browne returned to the Arctic on the *Resolute* in May 1850, as part of the Austin expedition, none of his visual work from that voyage appears to have been re-used. During the winter, he painted the drop-scene for the theatre aboard ship and was praised for his talents in the on-board periodical.[71] Eleven paintings and drawings by Browne from the Austin expedition exist in public archives.[72]

Transforming Geography: Regions and Landscapes

Despite its alleged realism, the panorama attempted to downplay the geographical aspects of the expedition in favour of creating more incredible spaces. As the booklet explained, it was divided into 'two distinct subjects, one-half the great circle exhibiting the Polar seas at midnight in the summer season, the other presenting a similar scene at noon, under all the sublime severities of an arctic winter'.[73] The accompanying booklet was available at the panorama at a cost of sixpence, and its sole illustration, a diagrammatic key (Figure 4.1), shows the content and layout of both halves of the vast painting. These two halves were separated by a partition (and possibly a curtain) on the viewing platform so that the views could be seen independently.

The title of the panorama, *Summer and Winter Views of the Polar Regions*, is the initial indication that geography will not take precedence here. The reader is placed firstly within seasons – times rather than places – and secondly within 'Polar Regions', a term so vague that it could apply to anywhere within the Arctic and Antarctic Circles, marking it out as a vast ill-defined realm to be traversed rather than lived in. The term 'Polar', more climatological than locational, is distinct from place names such as Spitsbergen or Boothia, which were included in titles of earlier Arctic panoramas at Leicester Square in 1820 and 1834. This title also contrasts with the titles of Burford's other panoramas such as those of *Sebastopol* (1855), the *Bernese Alps* (1853), and *Granada and the Alhambra* (1854). Burford's decision not to use a place-specific title becomes all the more significant considering the subject matter is highly geographical in nature. Unlike the expedition itself, the title does not locate us near Upernavik, Greenland, in summer or at the north-east tip of Somerset Island in winter, or even in the Arctic, but cultivates spatial uncertainty and, along with its composition, wishes to place the viewer in a time, rather than a place, into seasons, made distinctive by their light and dark variations. While the accompanying booklet mentioned geographical information, such as place names and coordinates, this was inundated by a sea of evocative description: 'desolation', 'wild disorder', and 'dreary shores' all combined in this 'sublime and splendid exhibition of icy grandeur'.[74]

Certainly, the locational information provided was lost on Mr Booley, a fictitious character who gave an account of his travels by panorama in Dickens's weekly journal *Household Words.* The adventurer, while mentioning Port Leopold, does not appear to be aware that a different location, over a thousand kilometres to the east, is represented in the summer view, or that the summer view was chronologically earlier than the winter:

> Mr BOOLEY slowly glided on into the summer season. And now, at midnight, all was bright and shining . . . Masses of ice, floating hither and thither, menaced the hardy voyagers with destruction . . . But below those ships was clear sea-water, now; the fortifying walls were gone; . . . and the sails, bursting from the masts, like foliage . . . spread themselves to the wind, and wafted the travellers away.[75]

The summer view shows the expedition 'in the month of July, in what was named Glacier Harbour, on the coast of Greenland, in latitude 73° 42′ N., longitude 55° 20′ W'.[76] Here, Burford gives the coordinates, nominally conferring an attribute of place on Glacier Bay. But the scene lacks any sense of earthly location; there is no attachment conveyed where 'all is wild disorder' and 'of which there exists no parallel'.[77] The sharp, peaked icebergs jostling with the ships indicate that the theatrical drama overrides geography and science. Burford's view has stretched the truth both visually and textually by creating a wildly romanticised world.

While the term 'Polar Regions' defines the space by climate rather than specific geographical places, the two halves of the panorama again mark the territory's space by meteorological effect: light and darkness. By emphasising the peculiarities of the extreme latitudes, Burford converts the Arctic into a space that becomes easily associated with the Burkean sublime, which found its source in many factors such as 'quick transitions from light to dark'.[78] The extremes of light and dark depicted here are in reality separated by six months and, as some expedition members noted in their narratives or journals, the darkness of midwinter was an experience that one became accustomed to gradually during the preceding months of increasing darkness. In fact, some light is visible during the Arctic winter even when the sun is eighteen degrees below the horizon. A member of the subsequent Austin expedition in 1850 observed: 'the daily decrease of light gradually habituates the mind to the unnatural change'.[79] On passing his first Arctic winter in 1853, Edward Belcher noted: 'This will close the month of January: not much unlike a gloomy English November, but not at all realizing the very cheerless long winter nights which have so frequently dinned our ears.'[80] In Burford's panorama, however, viewers stepped between summer and winter, light and dark, visually separated by a partition, in a matter of seconds.

Reviewers were particularly impressed with the representation of ice, particularly with its appearance in the summer scene, which showed the 'awful majesty of the Polar seas at summer midnight'.[81] The *Art Journal* noted the 'extraordinary and fantastic forms assumed by the icebergs'.[82] For the *Literary Gazette*, the icebergs were 'terrible and fantastic'.[83] The *Observer* noted that 'No tree, or corn, or verdure of any kind is visible to the eye, even in the season of summer . . . Never did the eye of man gaze on a more sublime scene.'[84] A reviewer in the *Gentleman's Magazine* felt that both sides of the panorama exhibited 'the fields of perpetual ice in their opposite aspects of the summer and winter seasons'.[85]

Burford's text and painting worked together to create a threatening and formidable world where the expedition was pitted against the environment. The peaked icebergs of Burford's summer view were in fact stylised and romanticised. His gothic renderings bore little resemblance to Browne's sketches, despite reviewers having been convinced of their reality. The diagram below (Figure 4.2) shows the evolution of Browne's painting through multiple media forms, from the rough painting he made in the

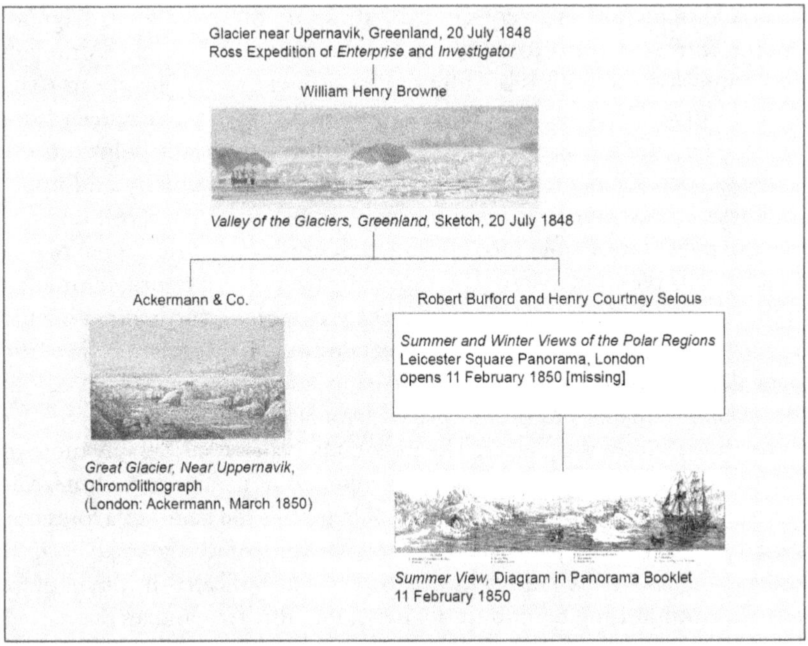

Figure 4.2 Diagram showing the evolution of *Valley of the Glaciers, Greenland* through multiple productions, 1848–50.

Arctic through to the lithography, the panorama itself, and the accompanying booklet.

The small painted sketch *Valley of the Glaciers, Greenland*, measuring only 8.3 × 30.5 cm and signed with Browne's initials, is dated to 1848 and, as will be explained below using topographical evidence, is almost certainly one early prototype for the summer half of Burford's panorama.[86] *Valley of the Glaciers, Greenland* (Figure 4.3) with its hastily laid-down wash and white gouache, was possibly painted on 20 July 1848, during the early part of the voyage, when an 'immense glacier was here observed extending completely along the imaginary coast line, on a level with the steep and elevated land ... The evening was cloudy but perfectly still.'[87] An additional representation of the glacier appeared as a chromolithographed plate in the folio, *Ten Coloured Views* (1850), which was published as the panorama opened.

The alteration of geography by Burford is here seen clearly in an examination of both views together (Figure 4.1 (top) and Figure 4.3), which allows the similarities and the differences to become evident. The view of a large glacier meeting the sea and the unusual proportions of the watercolour picture are very close to those of the panorama diagram, lending further support to the claim for the panorama's origin. Although the composition in both is broadly similar, Browne's version appears to be taken from sea level while Burford's panorama suggests a view seen from above. The numbers used in the key to identify significant points on the landscape and incidents lend the scene a layer of authenticity and imply accuracy. The ice may be the dominating feature of both compositions, but the nunataks[88] are visible too, if to a much lesser extent in the panorama booklet. The two nunataks of the panorama (centre and right) are situated in the same part of the composition as the watercolour, further making the case for the argument that the panorama derived its information from the scene depicted by Browne. On the left-hand side of the panorama version, the missing nunatak has been replaced by a large iceberg that reaches far above the height of the glacier. The panorama makes considerably more of the ice and reduces the land. Browne's rounded nunataks in the watercolour reassure the viewer of the presence of land, but the panorama loses this sense of stability, reducing the land to two comparatively small, jagged peaks rising above the glacier in the distance, as indicated in the booklet key. This minimising of land in the panorama further removes the viewer from the possibility of any solid, secure connection.

The lithograph *Great Glacier, Near Uppernavik*[89] (Figure 4.4) emerges as another link in the chain of Arctic reiterations of a single scene, and it may

Figure 4.3 William Henry Browne, *Valley of the Glaciers, Greenland*, 1848. Watercolour, 8.3 × 30.5 cm. © National Maritime Museum, Greenwich, London.

Figure 4.4 Detail from William Henry Browne, *Great Glacier, Near Uppernavik*, 1850. Chromolithograph, 14.5 × 24 cm. Royal Collection Trust / © Her Majesty Queen Elizabeth II 2021.

also have derived from the expedition watercolour *Valley of the Glaciers, Greenland* or a similar composition by Browne.[90] The lithograph is compositionally similar to the watercolour and panorama, again showing the face of the large glacier meeting the sea. Like the watercolour, the three nunatak summits are seen jutting out of the ice at the sides and in the centre of the glacier (left, centre, and right).

Taking the three images together, it is easy to see how the largest and most public depiction was the one that was farthest removed from its visual origins. From the expedition watercolour to the lithograph to the panorama, the ice grows from rounded, unthreatening forms in the earliest version into angular structures in the lithograph and finally, in the panorama booklet, to an extraordinary succession of Gothic structures that invoke the romantic sublime. The icy spires of the panorama pierce the horizon-line dramatically, their serrated forms erratic and restless, unlike the solid constant bulk of glacial ice represented in the watercolour. Not only that, but mirroring the spires on the left are the masts of the *Enterprise* likewise piercing the skyline (so tall that the mast-tops leave the page), suggesting that they are comparable opponents (this technique is repeated in the winter view to enable viewers to see the vast size of the ships).

The emphasis on the impressive nature of the ships, where icebergs and ships jostle for dominance, aligns with the language of conflict often used in describing the Arctic around the period of the Franklin searches. For example, a description of a visit to the ships of the Austin expedition, while docked at Woolwich in April 1850, described the *Pioneer* as a 'floating fortress against the fierce assaults of the Giant Frost' with enough stores to withstand the 'beleaguering siege of – it may be – a two or three years' Arctic winter'.[91] An article on Arctic exploration in 1860 in the *New Monthly Magazine*, written when the fate of the Franklin expedition was known, opened with the metaphor of war: 'The battle of the Arctic regions has been fought out and out again. On the one side is man, by nature weak, sensitive, and frail; on the other, privation, gloom, and cold, stern and ever-enduring.'[92] In Browne's watercolour the ships, while small and unobtrusive, provide a scale by which to gain some idea of the immensity of the glacier. Despite their small size, the ships do not appear to be in any danger.

The rounded masses of ice in Browne's watercolour may dwarf the two ships but they are far removed from the Gothic icy pinnacles of Burford's panorama, even as seen in the booklet, which was the schematic version of the original panorama. The text in Burford's booklet heightened the intensity of the summer scene by using evocative language that includes the stock descriptors like 'dreary' and the Gothic 'frowning', which had become associated with the Arctic in the public imagination:

> Desolation here reigns triumphant; all is wild disorder. The sea, piled into solid mountains of ice, strangely mingles its white pinnacles with the dark and frowning summits of rock that here and there rise to an immense height; and the earth, buried beneath its cumbrous load of frozen water, blends its dreary shores, undistinguishable by any boundaries, with the bleak deserts of the ocean.[93]

The description reads like the opening of a late eighteenth-century Gothic novel and conjures up a landscape of terror: a heavy, dark, and tumultuous world almost apocalyptic in nature in spite of the summer sun. The text also has echoes of the idea of a 'geological apocalypse' on the scale of Thomas Burnet's late seventeenth-century *Telluris Theoria Sacra* ('The Scared Theory of the Earth'), which imagined a huge deluge, a primordial drama, had caused the scars on what had once been a featureless globe.[94] In the panorama, the sea, with its mountainous peaks of ice, replaces the mountains.

The winter portion further reconstructs the Arctic, removing the reader from geography and into a more sublime space; the booklet describes 'two noble capes' showing some 'great convulsion of nature' that look down with 'dark frowning masses'. The descriptions of the mountains, violent, anthropomorphised, and disapproving, again recalling the seventeenth-century geological theories of Thomas Burnet, suggest turmoil and upheaval. 'Around in every direction the distance is one interminable waste, and desolate region of eternal winter.'[95] Simultaneously, the 'great convulsion' and the 'eternal winter' suggest both the beginning and the end of the world. The space is described in negative emotional terms, again invoking apocalyptic imagery, implying that here is a space of nothingness, which is alienating and incompatible with life. This vast region, then, is a non-place; it is 'interminable', tilting as far as one can imagine away from the sun, and out into space itself.

Enhancing Gender: Masculinity, Activity, and the Challenge of 'Man versus Nature'

The Arctic of Burford and Selous is also an icescape peopled by ships' crews. Although Inuit are not present, their absence is not necessarily misleading; the winter quarters of the Ross expedition were remote from Indigenous communities, and the large glacier depicted in the summer portion is around fifty kilometres inshore from the small island settlement of Upernavik. The figures we do see in the panorama are active and engaged in the outdoor pursuits of hunting, rowing, trapping, and building. These icescapes, populated by physically active figures, are in stark contrast to primary visual material from Browne, which often shows standing or seated figures in repose. The active bodies connect with the perception of the Arctic environment as 'threatening and unstable, a place not for reflection but for action'.[96]

As Andrew Lambert notes, some agents were keen to play down the scientific work that was carried out on the searches, showing that this was a purely heroic mission to save their fellow countrymen and not an excuse to carry out scientific observations, to complete the Northwest Passage, or to lay claim to the territory.[97] Garrison points out that these activities, which are found throughout the panorama, 'may also have suggested indirectly that much activity was taking place in the Arctic. Much activity, that is, in search of Franklin, undermining suspicions that rescue attempts were simply being appended onto voyages whose real priority was to cross the Northwest Passage.'[98] The active figures, labouring in the almost-theatrical spaces, also strongly suggest the 'man versus nature' trope. Kirsten Hvenegård-Lassen observes that the 'humanity versus nature' storyline performed in twenty-first-century scientific Arctic narratives includes 'tropes associated with heroic polar explorer myths of the past', such as 'heroic endurance' and the 'human struggle for survival'.[99] Indeed, one of characteristics of the colonial narrative of the Arctic is that 'nature becomes an agent in the story', and 'man's struggle against the elements' is a key component of the narrative.[100] Both halves of the panorama displayed overtly masculine activities associated with Arctic exploration, yet none of these activities is depicted in Browne's extant work.

Although Burford's depiction of the ships overwintering in Port Leopold, Somerset Island, or the *Winter View*, was derived from a sketch or painting that does not appear to survive,[101] other visual sources linked to Browne's primary visual depiction and to the panorama exist. The figure below explains the complexity of the visual sources, showing how Browne's initial painting or sketch inspired an engraving, a lithograph, and a panorama (Figure 4.5). An engraving of the panorama was then made as a visual key, and a section of it was engraved for the *Illustrated London News*. The earliest representation to be published was an engraving captioned *The Expedition Housed in for the Winter*, which appeared in the *Illustrated London News* only five days after the ships returned (Figure 4.6).[102] The accompanying article, which used three unattributed

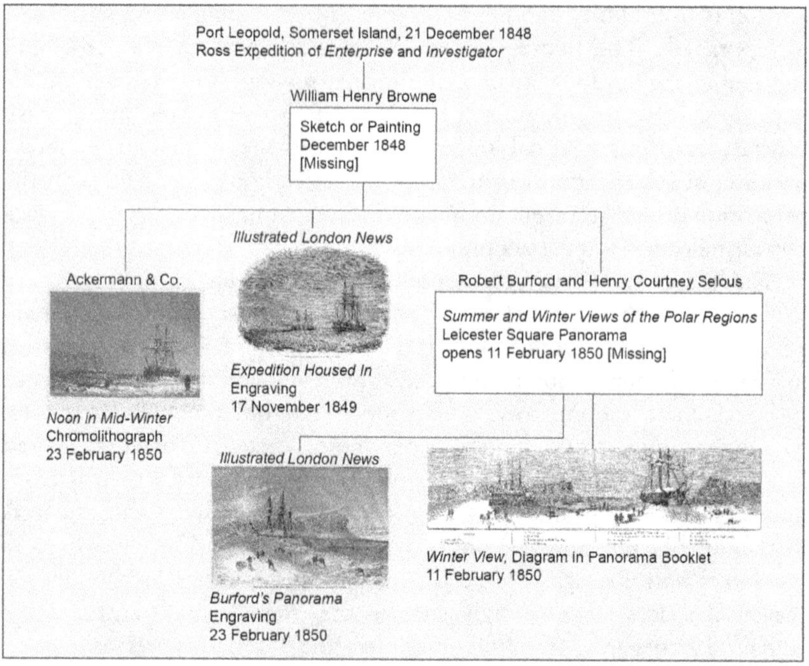

Figure 4.5 Diagram showing the evolution of *Winter View*, through multiple productions.

engravings to illustrate the news story, stated that they had been 'fortunate enough to succeed in obtaining some Sketches of the scenes of peril to which Sir James Ross, has been exposed'.[103] Although the engraving shares a subject matter and an aspect (or viewpoint) with the scene represented in the winter portion of the panorama, the former is unpopulated and sombre, very similar to a lithograph, *Noon in Mid-Winter* (Figure 4.7), which was published as part of the set *Ten Coloured Views* in February 1850. In *Noon in Mid-Winter*, however, three stationary figures inhabit the scene, and the use of colour allows the depiction of the sun below the horizon.

By contrast, the winter portion of the panorama, as shown in the booklet (Figure 4.1 (bottom)), presented a heavily peopled drama amidst the strange environment.[104] This space is the scene in which the expedition members perform their masculinity undeterred by the 'eternal winter'. The scene, showing officers and men engaged in almost every conceivable outdoor activity, is unparalleled in Browne's surviving work, which often shows a landscape, occasionally foregrounded with two figures.

THE EXPEDITION HOUSED IN FOR THE WINTER.

Figure 4.6 William Henry Browne, *The Expedition Housed in for the Winter*, 17 November 1849. Engraving, 9.5 × 14.5 cm. © Illustrated London News / Mary Evans Picture Library.

Figure 4.7 William Henry Browne, *Noon in Mid-Winter*, 1850. Chromolithograph, 16 × 23.5 cm. Royal Collection Trust / © Her Majesty Queen Elizabeth II 2021.

While the panorama booklet shows the exact layout of the visual elements in the exhibition itself, an engraving in the *Illustrated London News* (Figure 4.8), which appeared not long after the opening of the panorama, provides perhaps the most vivid representation of its winter portion.[105] Despite the monochromatic constraints, the engraver has skilfully conveyed the luminosity and dramatic effects of the panorama. The large illustration covers almost half a page and takes immediate precedence over the text.

The image of 'masculine' work displayed in Burford's panorama, so clearly shown by the *Illustrated London News* engraver (Figure 4.8) and in the panorama booklet, contrasts sharply with the domestic role of the ship as a home when wintering over, a home that included sewing and laundry duties, tasks more typically associated with women's work. As Heather Davis-Fisch has noted, the space of the ship was a domesticated one.[106] Although Ross's log for the *Enterprise* during this period records activities such as building the observatory, trapping foxes, and building the snow wall, these are described over a period of several months, whereas in the panorama the activity is clustered together simultaneously, creating a busy scene. In the background, barely visible, men pull a sledge through the darkness, resulting in a picture that includes a total of thirty human figures. The panorama contrasts sharply with the image that first-hand accounts often provide, which includes ample time for reading, drawing, and the necessity of keeping house. For example, John Matthews, who wintered in the Arctic for six years aboard the *Plover* (1848–54) described in his journal how he washed blankets on the ice in a tub with soap and boiling water: 'Indeed, Mother and Sister both, I may say that if you only saw me you would willingly hire me for your monthly wash.'[107] William Chimmo, aboard the *Herald* (the *Plover*'s supply ship), recounts at length in his narrative the manner by which costumes, including those for an 'elderly lady', a 'servant maid', and a 'bride', for a theatre performance were created: 'The midshipman's berth was like a dressmaker's shop! *All* were employed, even those who could but "sew on a button" . . . while those who had the advantage of sisters had learned to go through the more critical part of cutting out *the* dress.'[108]

Sherard Osborn described a typical day in winter on the subsequent Austin expedition of 1850 to 1851 in his published narrative: 'knots of two or three would, if there was not a strong gale blowing, be seen taking exercise at a distance from the vessels; and others, strolling under the lee, discussed the past and prophesied as to the future . . . If it was a school night, the voluntary pupils went to their tasks, the masters to their posts; reading men

Figure 4.8 *Burford's Panorama of the Polar Regions – the 'Investigator' Snow-walled in for the Winter*, 23 February 1850. Engraving, 17 × 27 cm. © Illustrated London News / Mary Evans Picture Library.

producing their books, writing men their desks, artists painted by candle-light, and cards, chess, or draughts, combined with conversation, . . . served to bring round bed-time again.'[109] The outdoor activity that Osborn described is leisurely, sociable, and done to break 'monotony', which was 'the only disagreeable part of our wintering'.[110] His references include the exercise that could be taken when the weather was fine and the opportunities to learn, paint, read and write, or play games during the winter period. His words are a far remove from the frenzy of activity depicted in the panorama's winter portion, as rendered in the engraving of the panorama by the *Illustrated London News*.

The summer half of the panorama also displayed a more overtly masculine Arctic than that of Browne. While the birds in flight in Browne's original watercolour (Figure 4.3) signify life, they also give a tranquillity to the scene; in the panorama, this contemplative image has been replaced with the presence of a bear being hunted, drawing attention to an activity that is both masculine and novel.

Neither does the lithograph *Great Glacier, Near Uppernavik* (Figure 4.4), intended for a more elite and scientific audience, show the active men who populate the panorama, but two restful and contemplative figures – the artist and his companion – who are calm and tranquil, relaxed in their environment, embodying a sense of place through their recording of the surrounding environment. In the lithograph, a rock outcrop is included in the foreground, providing a platform where the individuals are at ease. These figures, who here observe the glacier, can be seen in one of Browne's three surviving watercolours from the same expedition, *Coast of N. Somerset* (1849), in the Canadian archipelago.[111] Although a different location is delineated in this Arctic sketch, the two relaxed figures, one seated and one standing, appear. These figures too are less overtly active than their counterparts in the panorama, for they are scientific observers who have no need to do battle with the Arctic. Their purpose is to observe and record, while in the picture itself they provide a sense of scale.

The overt display of masculine activity in the panorama suggests characteristics of a moral sublime such as 'intrepid fortitude' and a disregard for 'every feeling arising from the consideration of himself'[112] as they endeavoured to find Franklin in a 'trackless waste of everlasting ice and snow'.[113] The booklet accompanying the panorama described how the search expedition had to 'grapple with difficulties of no ordinary nature' and 'endure toil and privation, and the perilous incidents ... which, by skill, daring, and steady perseverance, they triumphantly surmounted. The whole enterprise was nobly and gallantly conducted.'[114] This text complements the visual depiction of the Ross expedition through the panorama, which sends a clear message of masculinity that was far removed from the visual records created by the expedition members like Browne. As Catherine Lanone has pointed out, 'the discourse of polar exploration was a gendered discourse, casting men as heroic discoverers and women as compliant admirers endlessly waiting for their return – and their stories'.[115]

In fact, it was not uncommon for women to accompany their husbands on long whaling voyages, which included those that ventured into Arctic waters. It is also recorded that a single 'Englishwoman' took part in the search for Franklin aboard a private yacht, the *Nancy Dawson*, in 1849. Referred to as the 'heroine of Point Barrow' in a narrative by an anonymous midshipman not published until 1860, her exploits do not seem to have been known to the general public in the early 1850s.[116] Her presence at theatre performances and parties was remarked upon on the *Herald* and the *Plover* where 'her society was much sought after', and she joined in the singing with a rendition of 'Love's Young Dream' and 'Erin my Country'.[117] Another member of the *Plover*'s crew noted privately: 'I must here mention that he [Robert Sheddon, the captain of the *Nancy*

Dawson] had an Englishwoman with him who will gain a little celebrity by her voyage being the only one who has ever been in those latitudes before her name I believe is Emily.'[118]

Sublime Effects: Staging the Aurora Borealis

Added to the desolate winter scene was a meteorological phenomenon that heightened the intensity of the public's experience. Burford's *Summer and Winter Views* included the depiction of the aurora borealis, or northern lights, which vividly illuminated the winter portion of the panorama. Its appearance here emphasises the use of the Arctic and of the canvas as a dramatic space in which novel meteorological events are staged. The presence of the mysterious aurora was a significant way in which the places the expeditions lived in were transformed into supernatural and sublime spaces, in order to attract attention, draw visitors, and maximise profits amidst the multitude of attractions available at Leicester Square. Even a short announcement for the panorama's opening in the *Illustrated London News* found space to remark upon the 'sublime effects of an Aurora Borealis' that appeared in *Summer and Winter Views*.[119]

The aforementioned Mr Booley also noted the 'vivid Aurora Borealis . . . by night or day',[120] while for William Thackeray's fictional, satirical letter-writer Goliah Muff in *Punch*, the 'livid northern lights, the killing glitter of the stars; the wretched mariners groping around in the snow' were so alarming that he 'would not allow [his] children to witness it'.[121] The *Literary Gazette* felt that the presence of sunshine and the aurora in the summer and winter portions respectively served 'only as if to mock the sterility and utter coldness of the world'.[122] For the writer in the *Observer*, the winter view was 'appalling'. 'A scene of more perfect desolation than that which is here presented to the eye could not be imagined.'[123] In the *Era*, the reviewer noted that 'the phenomenon of the Aurora Borealis is very cleverly treated by the artist; it is represented as it appears towards the magnetic pole, in brilliant coruscations of every prismatic colour . . . we much question if these mimic regions will not be visited by all who take any interest in the fate of Sir John Franklin.'[124]

A key way, however, in which the panorama, both visually and in its descriptive text, differed from what we know of Browne's visual experience of winter in Port Leopold was in the depiction of the aurora borealis. Taking the engraving that appeared in the *Illustrated London News* of November 1849 (Figure 4.6) and the lithograph *Noon in Mid-Winter* (Figure 4.7) as the closest thing we have to Browne's primary visual record of winter quarters, we observe that the headlands in both renderings are in roughly the same places and the subject matter is ostensibly the same. The most obvious difference is the complete absence of the aurora borealis in the earlier engraving and in the lithograph. The engraving is devoid of any activity, either meteorological or human. Not a single figure is to be seen, and yet it is a calm and contemplative image showing the relative safety of ships wintering over in darkness. The lithograph *Noon in Mid-Winter* differs again. Here a single stationary figure observes the glow of the midwinter sun visible on the horizon, while two less-distinct human forms appear closer to the ship. The title, *Noon in Mid-Winter*, emphasises the strangeness of the winter darkness, but the absence of the aurora in both is notable in comparison to the panorama.

By contrast, Burford's booklet explains: 'Towards the south the hemisphere is splendidly illuminated by that extraordinary and beautiful phenomenon, the Aurora Borealis, – vividly darting its brilliant corruscations towards the zenith, and tinging the snow with its pale mellow light.'[125] The language suggests features of the Burkean sublime such as suddenness, dramatic transitions between dark and light, and extreme height. Later, in more detail, Burford repeated the phrase 'brilliant corruscations of every prismatic colour'. The reader is informed that the phenomenon is 'during the winter months almost constantly seen'.[126]

The *Illustrated London News*, which included an engraving of part of the winter half of the panorama (Figure 4.8), also drew attention to the aurora in its picture: 'The right-hand horizon glows with the splendour of an Aurora Borealis.'[127] The rays of light that represent the phenomenon here in the *Illustrated London News* are far more brilliant, dramatic, and obvious than in the booklet, reminding us that the booklet was intended to act as a guide during a viewing of the panorama, and not as a replica of the panorama in miniature.

The reviewer in the *Illustrated London News* lauded the overall effect of the panorama, concluding: 'The picture is painted throughout with wonderful power and intensity of effect, characteristic of the supernatural aspects of the Polar Regions.'[128] This reference to the supernatural complements the sublime and indicates the powerful attraction of mystery in the nineteenth century, something that still shrouded the polar regions. The uncertain scientific understanding of the aurora at the time meant that the phenomenon was still an object of some mystery,[129] although it had been established in the early part of the nineteenth century that it occurred only in certain latitudinal zones.[130] Many questions regarding the nature of the aurora, its origins, and appearance were still being debated around the time of the Franklin searches, and magnetic properties were an important facet of its study.[131] A connection between the aurora and magnetism had been identified in the early nineteenth century.[132] Geomagnetism (then known as terrestrial magnetism) was an important science, and the search expeditions from Britain were provided with equipment to take readings in the vicinity of the magnetic north pole.[133] The interaction of solar winds with the earth's magnetic field produce the aurora, thus making geomagnetism and the occurrence of the aurora inextricably linked. It was standard practice to record auroral activity among expeditions sent to the Arctic during the nineteenth century, and *The Admiralty Manual of Scientific Enquiry* recommended that auroral phenomena 'should be minutely registered, and all their phases, especially the formation, extent, situation, movement, and disappearance of arches, or any definite patches or banks of light'.[134]

Although the aurora formed a central component and an attractive part of Burford's composition, it fails to appear in any of Browne's surviving visual records from the Arctic. Similarly, I have not come across depictions of it in the primary visual records from the Franklin search expeditions, unlike the phenomenon of the paraselene, of which several sketches exist. One could argue that the aurora was difficult to paint (which it undoubtedly was) and that it would have to be done from memory due to intense cold and winter darkness. However, the reason for its distinct absence in the visual records becomes clearer when reading the ship's log of the *Enterprise* that winter.[135] The log recorded daily events or routines and meteorological information including auroral activity and shows that, far from the phenomenon being a regular occurrence, many days went by without it being sighted. When it does appear in the logbook it is often described as a 'faint aurora seen in the S to SE'.[136] Only on one occasion is it recorded as being 'brilliant',[137] using an adjective that comes nearest to the language employed in the public media. James Douglas Gilpin, clerk on the *Investigator*, wrote a serialised version of the expedition events for *Nautical Magazine* and described the aurora thus: 'The Aurora Borealis had shown itself frequently, but never in such splendour as I had expected to see it . . . In colour it was a light yellowish tinge . . . a pinkish hue.'[138] The clerk's expectations underline the association of the aurora more generally with high northern latitudes in the popular imagination. Although he suggests that the aurora was frequent, the logbook does not corroborate his statement. Significantly, Gilpin did not experience it in the 'splendour' he had expected, implying that widespread descriptions of the aurora as a specifically Arctic phenomenon in the nineteenth century leaned sharply towards hyperbole.

While Ross does not comment on the aurora in his official Admiralty report,[139] other commentators, on successive search expeditions in the same area, expressed some dismay at the lack of brilliancy and frequency of the aurora. Osborn, who travelled to the Arctic for the first time on the Austin expedition in 1850, remarked: 'With one portion of the phenomena of the North Sea, we were particularly disappointed – and this was the aurora. The colours, in all cases, were vastly inferior to those seen by us in far southern latitudes, a pale golden or straw colour being the prevailing hue; the most striking part of it was its apparent proximity to the earth.'[140] Belcher, commanding a five-ship search expedition in 1852–4, had a similar complaint:

> November 30. – About this period the season becomes extremely monotonous, and one is reduced to all kinds of imaginary reasons to account for the absence of expected phenomena, more especially the aurora ... I could hardly understand its prolonged absence. I had observed it, to the north of Behring's Strait, on the 25th August and continuously up to the 5th October, in its greatest brilliancy; and in Wales, at Swansea, in August.[141]

To Belcher's officers, John Cheyne and Walter May, the aurora became an elusive female whom they tried to capture with the magnetometer and electrometers.[142] Walter May also mentioned the delay in the aurora's appearance in his journal of the voyage, meaning that several months had passed during which it should have been possible to view the phenomenon. May's entry for 10 December 1852 noted: 'A good Aurora seen last night to the East. It is the first good one we have yet had in this country.'[143] The following night, the expedition again witnessed the aurora: 'We were favoured last night by a beautiful Aurora, It was first seen about 2.50 AM on the Southern horizon and gradually [neared?] the ship and eventually formed a complete arch over to the top of the hill on which the wires are placed, it was attracted by the wires, no doubt, for at one time it could not have been more than 100 feet from us – The magnetometer in the Observatory was affected but not the Electrometer.'[144] There is no mention here, then, of its sublime effects, although the aurora can be 'beautiful' and there is clearly an expectation of the display to be visually arresting. It was not only the searchers for Franklin who recorded disappointment, as earlier records from expeditions in the area of the Northwest Passage attest.[145] Ultimately, the written record shows that, far from being consistently seen in all parts of the Arctic, the aurora could be a more elusive event.

The explanation for the aurora's infrequent and unimpressive displays becomes clear when the locations of winter quarters of the expeditions in question are considered. The search expeditions that entered the archipelago via Baffin Bay were so far north that they were within the auroral oval that forms around the geomagnetic north pole, which also explains why it was seen in the southern skies. And yet somehow, by the time of the panorama *Summer and Winter Views*, the aurora had become so synonymous with the polar regions that it was unthinkable not to include it when a large public exhibition was created. The aurora borealis had become a 'crucial component' in the construction of the Arctic imaginary.[146]

Conclusion

The panorama was such a convincing illusion that people regarded it as a realistic source of information with the ability to bring them closer to the very place where the Franklin expedition might be. The reviewer in the *Era* considered that, given the concern for the missing expedition, 'anything tending to enlarge the sphere of thought, or add to the scanty stock of information relative to the trackless waste of everlasting ice and snow, is sought for with the utmost avidity'.[147] Lady Franklin herself found the spectacle so compelling that she 'remained two hours inspecting the picture, which must possess a peculiar interest to her, as being near the place in which her husband and his expedition are supposed to be, if still alive'.[148]

The overall effect of Burford's panorama was to displace the viewer in time and space, transporting them not to the Arctic but to a supernatural space in which the expedition's sense of place was lost in the overwhelming aspect of a sublime expanse of icy pinnacles, light and dark, tumultuous upheaval, and masculine endeavour, combining a sublime that was at once gothic, mysterious, and supernatural. By closely examining the panorama based on William Browne's work, I have shown how the visual record of an expedition could be transformed for popular consumption and that the panorama *Summer and Winter Views*, alleged to imitate reality, differed vastly from the Arctic visual experience of William Browne. The gothicisation of the icescape through the panorama had the effect of positing the search ships as well-matched foes of Arctic ice.

The public representation of the Arctic in the 1850s, as actions, icescapes, and phenomena were enhanced, provided a theatrical stage on which to demonstrate the gallant efforts of the search parties. While Browne's figures in sketches and lithographs often epitomise calm, rational scientific observation, those in the panorama engage in more active, overtly masculine endeavour. The very public panorama emphasised the moral sublime in action and the fervour of activity that was being undertaken in the search for Franklin. It showed movement and mobility through space rather than stillness in place. The result was that the Arctic became less of a place in the geographic sense and more a space in which the moral sublime could be effectively displayed. The experience of viewing the panorama, visited by a broad cross-section of society, must have left an overwhelming impression on the minds of those who saw it. It is possible that even naval officers were misled by its vivid depiction of the aurora, as records of disappointment concerning the aurora are found in Arctic narratives. The wide appeal of the panorama, and the affordable cost of admission, meant that this theatrical spectacle, showcasing British bravery amidst 'savage horrors', had the power to 'attract the public as the needle to the pole'.[149]

CHAPTER 5

'Power and Truth'

The Authority of Lithography (1850–1855)

By April 1854, Samuel Gurney Cresswell was on board the *Archer* off Elsinore in Denmark as part of the Baltic fleet.[1] Back in London, sketches that he had created were being transformed into a series of eight lithographs that would be published in colour by Day and Ackermann.[2] It was not until 31 July that he wrote to his parents from the *Archer*: 'I am very glad to hear that the sketches are done so well, pray send me a copy out.'[3] Cresswell's letter to his parents serves to remind us that naval officers were primarily in the service of the navy and were not overseeing the publication of prints derived from their on-the-spot drawings. Although Cresswell had no input into the final appearance of the lithographs, which were very different in style to his sketches and paintings, such published prints carried, and indeed still carry, an aura of authenticity due to their close associations with 'factual' visual records.

The concern with truth and accuracy that permeated the discourse surrounding Robert Burford's panorama extended to lithographs produced of the Arctic during the Franklin searches. Their format, price, and close association with naval officers, whose sketches they purported to reproduce, indicated an authority that could be 'relied upon as exactly truthful'.[4] These views, considered as artefacts replete with knowledge of the Arctic, displayed 'much power and truth'.[5] However, much like the panorama painters, the lithographic artists who worked on the Arctic prints used their imagination and aesthetic awareness to create a different version of the Arctic. As the search progressed, their representations heightened the sense of humanity's battle with nature through perilous positions and performative suffering. The convincing presentation of the lithographs still leads to their reception as authentic 'sketches' and factual records amongst critics and the public today.

Like the panoramas, sets of lithographs based on officers' drawings created new versions of the Arctic imaginary. Tensions between knowledge and 'truth', on the one hand, and a concern with portraying the difficulty of the search, on the other, become apparent through examining lithographs produced from the search expeditions. As a medium, lithography was associated with natural history and other sciences as well as with landscape views. However, the cost of lithographs put them well beyond the means of the average panorama-goer.[6] *Ten Coloured Views Taken during the Arctic Expedition* (1850) by William Henry Browne was the first such folio of lithographs from the Franklin search to be published.[7] It was followed by two more sets in 1854 and 1855: Cresswell's *A Series of Eight Sketches in Colour ... of the Voyage of H.M.S. 'Investigator'* (1854), which included a map of the newly discovered Northwest Passage, and Walter William May's *A Series of Fourteen Sketches Made during the Voyage up Wellington Channel* (1855).[8] A second edition of Cresswell's *Eight Sketches* was completed at Day and Son in January of 1855.[9] Edward Augustus Inglefield, who led three summer expeditions to the Arctic from 1852 to 1854, also had four tinted lithographs printed in late 1853 that showed 'critical' and 'perilous' situations of ships among the ice; these were published by Dickinson Brothers and sold individually.[10]

In this chapter, I look closely at a selection of these lithographs through case studies in transformation, dissemination, and reception, noting how an almost irrefutable authority was conferred on the prints that were done from on-the-spot sketches by officers on the Franklin searches. Browne's lithographs show the conflicting interests of scientific accuracy and commercial potential, the latter exemplified by the tendency of lithographers to heighten sublimity in order to increase the attractiveness of prints. Cresswell's folio reveals how an officer's uncertain sketches could be rendered as confident and apparently reliable prints by accomplished lithography artists. Finally, the lithographs by May elide any cheerfulness from his work, in favour of struggling figures that epitomise the idea of 'heroic failure', a concept that increasingly became attached to polar exploration from the period of the search expeditions.[11] In both Cresswell's and May's prints, and to a lesser extent in Browne's, the power of nature and the 'man versus nature' trope increasingly stand in as an alibi for failure.

Despite their frequent use to illustrate present-day popular texts on polar exploration, the lithographs themselves have received little sustained critical attention. Instead, they decorate publications and are frequently taken as factual artefacts, serving as eye-catching enhancements rather than as objects of critical inquiry. As Brian Maidment notes, 'prints, however documentary their mode or literal their intentions, can never be entirely naturalistic representations of the past'.[12] Bernard Smith reminds us too that documentary drawing, based on the perceptions of draughtsmen, also relies on 'a stock of visual memories drawn from their worlds'.[13] Nevertheless, it is likely that prints such as lithographs, when not exact copies of the original drawings, combine a greater proportion of inventive drawing from the mind of the lithographic artist. A common mistake among scholars is a failure to distinguish between lithographs – the product of printing houses in the metropole – and paintings or sketches done by expedition members in the Arctic. References to lithographs as 'paintings' or taking prints to be exact reproductions of an artist's work are frequent errors. Such analyses fail to take account of intermediaries: the work of the accomplished lithographic artists such as William Simpson, Edmund Walker, and Charles Haghe, the printing firms like Day and Son or Ackermann, which ultimately were commercial operations, and the influence of the Admiralty, to whom all records from expeditions had to be given. Lithographic artists generally took on the production and supervision of lithography in the metropole.[14] This was particularly the case where amateur artists (such as the Arctic officers) were concerned, although professional artists might draw on the lithographic stones themselves.[15] In fact, in the case of some of the Arctic lithographs, even a cursory glance shows the paintings and preparatory sketches to which we have access to be very different from the published lithographs. This suggests that the lithographic artists exercised their own imaginations and what they had learned from viewing earlier Arctic representations to enhance, improve, or exaggerate the officers' drawings. However, the use of the word 'sketches', in the titles of Cresswell's and May's lithograph folios, seeks to disassociate the works from any intermediary. The word implied 'authenticity and truthfulness', a certain 'truth status', and sketches were seen as establishing an 'intimate relationship between artist and audience'.[16]

This chapter analyses the content of the lithograph folios while paying close attention to the officers' sketches, the geographical contexts, and the specific origins of the lithographs, reminding us that, as Tim Youngs reiterates with respect to travel writing, 'we should not assume a uniformity of mode or perception of travel in the nineteenth century'.[17] Furthermore, the dissemination and reception are attended to by the examination of contemporary reviews, advertisements, and other information. Throughout the discussion that follows, I argue that, in some cases at least, the lithographic responses to the Arctic are more likely to be those of British lithographic artists, publishers, and Admiralty figures than of expedition members whose sketches they purported to reproduce.

Invented in the late eighteenth century, lithography is a planographic technique of printing on limestone that was valued particularly for the illustration of topography and natural history, as it facilitated finely graded shading and details that were difficult to achieve using engraving.[18] It was considered to be a more direct method of reproduction, since engravers were not required to transfer the image to a plate or block and drawings did not have to be converted to linear designs as they did for engravings.[19] The process was also considerably cheaper than engraving.[20] Louis Agassiz, who always employed his own artists to illustrate his work, set up his own lithographic printing operation in 1838 in order to improve standards of illustration in scientific texts including his own, and his plates, such as those in *Études sur les Glaciers* (1840), were admired by the scientific community. Chromolithography, the printing of each colour of the print using a different stone, was developed commercially from 1850.[21]

Although lithography was more economical than other printing methods, colour illustrations were still expensive, and the sets of Arctic lithographs ranged in price from Browne's folio at a cost of sixteen shillings, or twenty-one shillings ('handsomely bound'),[22] to Cresswell's larger *Series of Eight Sketches in Colour*, at a price of two pounds, two shillings for the set, more expensive than any written narrative to come out of the search expeditions.[23] Inglefield's large lithographs were offered for sale individually at one guinea (one pound and one shilling) each,[24] and May's *Fourteen Sketches* were twenty-one shillings (or one pound and one shilling) for the set.[25] To put this into perspective, in 1850, a seaman on a sailing ship earned on average forty-five shillings a month, and a printer in London earned thirty-six shillings a week.[26]

The lithographs varied dramatically in size as well as price; Browne's averaged about 14 × 20 cm, Inglefield's 40 × 70 cm, Cresswell's 40 × 60 cm, and May's 18 × 30 cm. Although we might assume that the relatively high prices indicate a more elite audience for the lithographs than many of the people who could afford the entry price of one shilling to see the panorama *Summer and Winter Views of the Polar Regions*, an article in the *Morning Chronicle* in 1855 noted that lithographs were 'constantly before the eye of the public' and 'almost completely fill the windows of our printsellers',[27] reminding us that they too were part of the Victorian culture of exhibition and were visible to all sectors of society, even if they did not have the means to purchase them. Indeed, when Inglefield's lithographs were being printed, advertisements in newspapers show that his 'celebrated pictures' of the Arctic were on view at Dickinson Brothers for several months.[28] The visibility of Arctic prints through their display in printers' windows and at their establishments indicates a much wider audience for these expensive representations than has generally been assumed.

William Henry Browne's *Ten Coloured Views* (1850)

Browne's *Ten Coloured Views* marked the beginning of illustrated publications from the search expeditions.[29] The work contained a 'Summary of the various Arctic expeditions in search of Sir John Franklin' to that date and ten plates of lithographs that represented 'faithful delineations of the most interesting scenery'[30] encountered on the expedition of the *Enterprise* and *Investigator* (1848–9) led by James Clark Ross. The list of plates and summary of expeditions were placed before the lithographs, which were captioned with titles. Although the plates were copied from Browne's work, he was not the author of the accompanying text.[31] This text was printed in both French and English; the former was regarded as the international language of science at the time, which would suggest that the folio was aimed at a scientific audience as well as those purely interested in the search. In fact, eight out of the ten plates show an interest in geological features at a time when little was known about the geology of the Canadian Arctic archipelago. The subsequent lithographs of Inglefield, Cresswell, and May were larger and more expensive and, as I will demonstrate later in the chapter, more interested in conveying peril and difficulty. The folio *Ten Coloured Views* was advertised in the *Athenaeum*, on 23 February 1850, less than a fortnight after Burford's panorama had opened, making the two closely linked by Browne's name and by their timing.[32] The advertisement noted too that they were 'drawn by Lieutenant W. H. Browne, late of Her Majesty's Ship "Enterprise"',[33] thereby attaching the authority of a reliable eyewitness to the production of the lithographs and implying that they were direct copies of Browne's work on that expedition

'Tenfold interest' attached to *Ten Coloured Views* on its publication 'from the uncertainty that hangs over the fate of a brave body of our countrymen, who may even now be living in the desolate regions here depicted'.[34] The *Athenaeum*, in its review of the lithographs, noted the 'extreme interest evinced by the public at the present moment in all that relates to Arctic Expeditions'.[35] The reviewer in the *Critic* concluded that the 'portfolio will be a most welcome addition to the drawing-room table. It is the novelty of the season.'[36] In common with *Summer and Winter Views*, reviewers were concerned with the accuracy of the representations, but some regarded the lithographs as being more reliable than the panorama. The *Critic* noted 'all who have seen the panorama in Leicester-Square, which was painted from the drawings of the same gentleman, will be eagerly desirous to possess these more probable reminiscences of scenes, which it is difficult for the most vivid imagination to paint'.[37] Although the panorama had been lauded for its truth and accuracy, this implies that some viewers still suspected that the deception of the panoramic format might extend to its content, or realised that Burford had used his imagination to transform Browne's sketches into the panorama. The same reviewer spoke of how *Ten Coloured Views* derived 'additional value' and could be 'relied upon as exactly truthful' since they were 'almost *fac similies* of sketches taken upon the spot by Lieut. Browne'.[38] This suggests the reviewer may, in fact, have seen Browne's sketches at Ackermann's. The *Art Journal* spoke of Browne's folio as a 'valuable addition to our knowledge of the Arctic regions' and considered the views to show 'much power and truth'.[39] The *Athenaeum* recommended it to those 'desirous of extending their knowledge of the wonders of the northern seas'.[40]

However, the reviewer in the *Athenaeum* remarked 'Great credit is due to Mr. Haghe for the fidelity and spirit with which he has lithographed Mr. Brown's [*sic*] drawings'[41] and the title page of *Ten Coloured Views* stated 'Drawn by W. H. Browne Esq., Lieut. R.N., Late of H.M.S. "Enterprise". On Stone by Charles Haghe.' These acknowledgements indicate that some people were aware of the importance of the role of the lithographer and that it was perhaps not a given that fidelity could be depended upon. Around the same time that the lithographs were being prepared, Browne was writing his letter, printed in *The Times*, the *Athenaeum*, and the *Literary Gazette*, regarding the use of his sketches for Burford's panorama.[42] The letter gave his address as being in Birkenhead, Cheshire, indicating that he did not reside in London and would not have been closely involved with the production of the lithographs.[43] The lithographs depicting the Franklin search were sometimes quite different from the officers' sketches. Artists at printing houses were probably encouraged to make the work of the lieutenants (who after all were amateur artists) more commercial and aesthetically pleasing. Often, this meant heightening the sublime or picturesque effects of a composition even though the lithograph folios, by their very format, suggested accuracy.[44] This effect was compounded by the direct association of representations with officers who had participated in the expedition and made on-the-spot sketches.

Browne's folio of lithographs displays significant scientific, romantic, and commercial registers (such as the lure of the sublime). This complexity results in an uneven adherence to scientific accuracy, although some of the lithographs function as geological records and simultaneously as representations of the sublime. Only three extant works exist by Browne from the Ross expedition (1848–9), and two of these bear some resemblance to two of the lithographs.[45] The titles of the lithographs, listed below, betray an interest in geology that seems to have influenced the selection of views to be represented from the expedition. A cursory glance shows that geographical keywords such as 'glacier', 'fiord', 'ravine', 'cape', and 'cliff' figure prominently.

***Ten Coloured Views*: List of Plates by William Henry Browne/Charles Haghe**

Great Glacier near Uppernavik, Greenland.
Fiord near Uppernavik.
Ravine near Port Leopold.
The Bivouac, Cape Seppings, Leopold Island.
North-East Cape of America, and Part of Leopold Island.
Termination of the Cliffs, near Whaler Point, Port Leopold.
Prince Regent's Inlet.
Remarkable Appearance in the Sky Always Opposite the Sun.
The Devil's Thumb.
Noon in Mid-Winter.

Fortunately for the lithographer Charles Haghe and the publishers Ackermann, the Ross expedition wintered at Port Leopold, on the north-east of Somerset Island, an area of high sea cliffs and ravines, meaning that the topography included features associated with the sublime and thus negated the need for extreme exaggeration on the part of the printers. The plateau of Somerset Island, which consists of limestone and sandstone, ranges in elevation from three to six hundred metres and is cut by steep-sided river valleys.[46] The majority of the ten lithographs depict scenes of the north-west coast of the island, where they spent a prolonged amount of time.

Geology was one of the sciences that had the potential to capture the public imagination, and constant debates over geological theories during the nineteenth century stimulated the public appetite for novelty and knowledge. Seminal texts, such as Charles Lyell's *Principles of Geology* (1830–3), Louis Agassiz's *Études sur les Glaciers* (1840), Hugh Miller's *The Old Red Sandstone* (1841), and Humboldt's *Cosmos* (1848) popularised geology in both public and scientific realms.[47] Certainly, the library of the *Assistance* in 1852 included *Cosmos* and *Principles of Geology* as well as Henry De la Beche's *Geological Manual* (1831), William Buckland's

Geology and Mineralogy (1836), and four other unspecified geological books.[48] It is likely that Browne had access to a similar selection of geological texts during the voyage of the *Enterprise* and the *Investigator* from 1848 to 1849.

William Henry Browne had spent his childhood on the rocky headland of Howth in Ireland; one could suggest that this environment was the very place that influenced his choice of subject matter, the representations of geological features, which are so often the focus of his Arctic sketches. The peninsula of Howth Head is north-east of Dublin city, connected to the mainland by a narrow strip of land, and many parts of the coast consist of exposed rock and cliffs. From the harbour master's house in which he grew up, he would have looked out on the island of Ireland's Eye, where bare rock of sandstone, quartzite, and greywacke is exposed and meets the sky in places.[49] The entire island 'displays some of the best exposures of Cambrian rocks on the east coast of Ireland'.[50]

Browne's sledge journal from his second expedition shows a good knowledge of geology and observational skills, as he correctly identified rock types that he encountered in 1851 on the Austin expedition. He fulfilled many of the criteria for a geological observer as laid out by Darwin in *The Admiralty Manual* and identified limestone, sandstone, conglomerate, and granite on his sledge journey, noting aspects such as possible coal deposits, stratification direction, the heights of headlands and cliffs, and the colours and textures of rock. He included a separate piece in his journal on the topography and geology of the area he traversed, whereas other sledge leaders who were also good artists, such as George F. McDougall and May, hardly mention the geology of the land they pass through.[51]

One of the most popular lithographs, *The Bivouac, Cape Seppings, Leopold Island* (Figure 5.1), represents a sledge party beneath the cliffs of Cape Seppings on Somerset Island. The human figures act as a scale, giving the viewer an idea of the size of the geological form, and the delineation of the cliffs demonstrates an understanding of the sedimentary geology. There is no similar sketch by Browne in the archive, but twenty-first-century photographs of the area show a formation remarkably similar to that depicted in the lithograph. An examination of topographical maps confirms that the cliffs at Cape Seppings rise to three hundred metres and are not actually vertically exaggerated to convey the sublime in the lithograph. Despite this, the sublimity of a large geological mass that reduced the search party to minuscule figures was not lost on the *Art Journal*, whose reviewer commented on the 'savage grandeur' and 'sublimity about some

Figure 5.1 William Henry Browne, *The Bivouac, Cape Seppings, Leopold Island*, 1850. Lithograph, 24.5 × 17 cm. Royal Collection Trust / © Her Majesty Queen Elizabeth II 2021.

Figure 5.2 William Henry Browne, *Prince Regent's Inlet*, 1850. Lithograph, 24.5 × 17 cm. Royal Collection Trust / © Her Majesty Queen Elizabeth II 2021.

of these scenes, of a very striking kind', singling out *The Bivouac* as an example.[52]

In the bottom-left corner of *The Bivouac*, two figures apart from the main group are seen, indicating the scientific observer/artist and his companion. Other figures have arrived and are resting, unloading sledges, gathering around a fire. Although the presence of the expedition members beneath the sheer cliffs registers the sublime, their activities at the place of rest are less arduous than hauling the sledges. Moreover, the two observers, who are at ease in the environment, present a contemplative stillness and not a battle with nature. Their presence suggests a Romantic influence, urging a reconnection with nature and an engaged interest in the environment. This is not a sublime in which 'terror' or 'pain' indicative of the Burkean sublime are explicitly represented. The attribute of 'bodily pain, in all the modes and degrees of labour, pain, anguish, torment ... productive of the sublime' is lacking.[53]

This Burkean sublime does creep into the lithograph *Prince Regent's Inlet* (Figure 5.2), which may be based on one of Browne's three extant works from the Ross expedition, *Coast of N. Somerset – Regent's Inlet* (Figure 5.3).[54] In the latter, two observers in the landscape fulfil a number of functions: they provide a scale for the geological features; they signify scientific authority implied by representing the act of observation and on-the-spot drawing; they act as witnesses, implying truth. Furthermore, the figures could be interpreted as the *Rückenfigur*, which sought to draw the viewer into the painting and into appreciation of nature,[55] familiar from the work of the German Romantic artist Caspar David Friedrich. This connection of science with romanticism, including its use of figurative language, was commonplace in the nineteenth century, particularly in the study of ice and glaciated landscapes.[56]

In contrast, the lithograph *Prince Regent's Inlet* (Figure 5.2) shows a very different scene.[57] At first glance, the watercolour and the lithograph appear to have little in common apart from the place name in their titles. On closer inspection, it becomes evident that the lithograph is compositionally almost a mirror image of the watercolour and may also present the view from a different angle. Noticeable too is the way in which enormous quantities of snow lie heaped up in the lithograph, in contrast to the relatively light layers of snow in the watercolour. The blue sky of the watercolour has disappeared, replaced by a stormy and ominous backdrop,

Figure 5.3 William Henry Browne, *Coast of N. Somerset – Regent's Inlet, near Cape Leopold*, 1849. Watercolour, 20.5 × 34.5 cm. Scott Polar Research Institute, University of Cambridge, with permission.

as have some details of geology. The tiny man-hauled sledges are starkly obvious in the right-hand corner of the lithograph, highlighted against the white snow. Although two figures still sit (bottom centre) and observe in the lithograph, their presence is almost lost in the shadow of the cliffs. Furthermore, the lithographer has made the geological features here seem larger, showing that a heightening of the sublime must have occurred in some of the plates. The tension here between scientific accuracy and commercial interests comes to the fore, as a more threatening Arctic comes into view.

Few of Browne's lithographs show the struggle that becomes more pronounced in the later lithographs of Cresswell and May, where the heroic self-sacrifice of the search is made clear. Certainly, the figures at ease in the environment in *Coast of N. Somerset – Regent's Inlet*, and in some of Browne's lithographs, would seem to suggest contemplation rather than action. Nowhere is this more evident that in the lithograph *Ravine near Port Leopold* (Figure 5.4). The seated figure on the left is offset by a reclining figure in the right-hand corner, who contemplates the scene in a Romantic

Figure 5.4 William Henry Browne, *Ravine near Port Leopold*, 1850. Lithograph, 12 × 17.5 cm. Royal Collection Trust / © Her Majesty Queen Elizabeth II 2021.

pose. The figure on the left, probably intended to be the artist, sits and observes; two figures enter the ravine itself, unencumbered by sledges or supplies; their objective appears to be daily exercise and exploration of their local environment. According to the *Critic*, *Ravine near Port Leopold* was a scene 'to which the Alps afford no parallel'.[58] This comment, however, reminds us of the association of the Alps and Romantic scenery; the reclining figure brings to mind the tourism for which the Alps became known in the late eighteenth century. Indeed, around the mid-nineteenth century, growing tourism and mountaineering in the Alps attracted observers who peered at climbers through telescopes from the comfort of hotels.[59] These spectators were particularly interested in dangerous ascents, where there was a possibility of watching the climbers confront death.[60] However, in *Ravine near Port Leopold*, there is no sense of imminent death or disaster in the relaxed figures that people its lower portions, and the contents of the representation suggest an incompatibility with the apprehension of the Burkean sublime.

The choice of subject matter in the lithographs was surely influenced by several factors: Browne's interest in geology fostered by his background; the location of winter quarters near the impressive sea cliffs of Somerset Island providing a local environment of immense geological forms for a large part of the expedition; a possible interest in Romantic landscape painting; and the publisher's and the Admiralty's choice of what sketches to produce as lithographs and how much to alter them. Browne's lithographs remain complex documents that illustrate simultaneous registers of scientific accuracy and lithographic exaggeration. While Browne's own Romantic response to the environment did not preclude his accurate recording of topography and geology, traces of this may be seen in the relaxed observers that appear in both his watercolour from the Ross expedition and the subsequent lithographs. These remnants, seen in the calmness of some of the figures that disrupt the portrayal of the 'inhospitable land',[61] still give a sense of Browne's 'conversation' with the land, despite the mediation of the lithographer Charles Haghe, the publisher Ackermann, and the Admiralty. However, the lithograph *Prince Regent's Inlet* (Figure 5.2) points towards what would become the more typical lithographic representation of the search, with labour, difficulty, and the immensity of nature emphasised in the prints of Inglefield, Cresswell, and May.

Samuel Gurney Cresswell's *Series of Eight Sketches in Colour* (1854)

Cresswell first served on the *Investigator* during the Ross voyage of 1848 to 1849 (the same voyage during which Browne was on the *Enterprise*) and subsequently as second lieutenant, again on the *Investigator*, as part of Robert McClure's expedition, which entered the Arctic via the Bering Strait in 1850 and did not return until 1854. The expedition sighted the last link of the Northwest Passage early on and spent three winters in the region of Banks Island, eventually abandoning the ship to the ice and crossing the remaining section of the Northwest Passage on foot and by sledge, in order to reach the ships of the Belcher expedition.[62] Although McClure and most of the crew were unable to reach England until October 1854, the news of the discovery and traversing of the Northwest Passage was brought to England by Cresswell himself after he arrived, with McClure's despatches, in Britain on Inglefield's supply ship the *Phoenix* in October 1853.[63]

Unlike Browne's lithographs, in the case of Cresswell, we are fortunate in having access to a large selection of his paintings created during the two Arctic expeditions in which he participated.[64] Thirty paintings and sketches exist by Cresswell that are dated to the first voyage, and twenty-three pictures date from the second *Investigator* voyage. Several of these latter pictures are directly comparable with some of the lithographs in *Eight Sketches*, showing stark differences between the sketches and their respective prints.

Reviewers were, once again, concerned with the accuracy of the representations. Referring to the published lithographs, the reviewer in the *Art Journal* considered that 'Lieut. Cresswell's masterly sketches are powerful aids in enabling us to form a tolerably accurate notion of the hazards of an Arctic expedition'.[65] The reviewer also found space to note how they were 'excellently lithographed' by William Simpson and Edmund Walker (whose names also appear on individual lithographs) and the printers Day and Son. All in all, they considered them 'a series of interesting views' as well as 'beautiful works of Art'.[66] A reviewer in the *Morning Chronicle*, on the occasion of the publication of the second edition the following year, considered that the 'whole of the views are exceedingly interesting, they are lithographed in the highest style of art, and their accuracy is vouched by the name of Lieutenant Cresswell, whose rare abilities as an artist we have had occasion more than once to acknowledge'.[67]

Conversely, the *Athenaeum* was less than complimentary in its review. By opening with the phrase 'These interesting sketches seem faithful, though not very artistic', the reviewer questioned not only the artistic merit but also the accuracy of the representations. The cautious choice of the word 'seem' casts some doubt on the prints' fidelity to nature. Furthermore, the reviewer lamented the 'heavy' colouring and the 'want of delicacy in detail, which is peculiarly felt in snow scenes, where the tints are so soft and evanescent'.[68] The review continued: 'The variety of surface is not conveyed to the eye, and the result is an appearance of inaccuracy which we are sure does not exist.'[69] The unusual and enigmatic remarks undermine the lithographs' intent as convincing representations and are intriguing in light of comparisons with Cresswell's sketches, which indicate that much of the detail of the scenery relied heavily on the imagination of the lithographers Simpson and Walker.

Cresswell himself revealed that he struggled with his drawing, suggesting that he did not think of himself as an artist; in July 1848, he wrote to his parents from the Whale Fish Islands, Greenland, lamenting that he was experiencing difficulty: 'I have taken great pains with my drawing since I have been away and the more I try to draw, the more I am sure I have no taste for it. I will do my best.'[70] Indeed, his work from both voyages shows a laboured stiffness (noticeable all the more for its absence in the work of Browne and May) that reveals his frustration and lack of confidence in drawing, particularly with the human figure and with perspective.

The eight prints show specific and significant events relating to discovery and danger: headlands, the lone ship in a 'critical position', a storm, a sledge party setting out on what would be the completion of the Northwest Passage, the intense labour of sledging. Their titles, listed below, focus on the actions of the expedition and the movements of the ship. In this way they tell a specific narrative of exploration, one that did not include representations of leisure or inactivity.[71]

A Series of Eight Sketches in Colour: List of Plates by Samuel Gurney Cresswell / William Simpson and Edmund Walker

I First Discovery of Land by H.M.S. Investigator, September 6th 1850.
II Bold Headland on Baring Island.
III H.M.S. Investigator in the Pack. October 8th 1850.
IV Critical Position of H.M.S. Investigator on the North-Coast of Baring Island. August 20th 1851.
V H.M.S. Investigator Running through a Narrow Channel in a Snow Storm between Grounded and Packed Ice. September 23rd 1851.
VI Melville Island from Banks Land.
VII Sledge Party Leaving Mercy Bay, under the Command of Lieutenant Gurney Cresswell. 15 April 1853.
VII Sledging over Hummocky Ice. April, 1853.

In contrast to the lithographs, one of Cresswell's paintings from the voyage, *Brown's Island, Coast of America* (Figure 5.5), shows men scattered about in a relaxed manner under a blue sky in August 1850.[72] The painting reflects the periods of inactivity that were inevitably typical of search expeditions. In the foreground of this painting, a figure reclines (bottom left) while, nearby, two sailors light their pipes; other groups of men stand around, close to the partly concealed ship. The scene has an air of quiet joviality and restfulness, not commonly seen in the public representations. The lithographs instead highlight discovery and difficulty, new lands, and the ship in peril. The first lithograph in the series immediately opens with a 'discovery'. *First Discovery of Land by H.M.S. Investigator, September 6th 1850* (40.5 × 55.9 cm), lithographed by Simpson, sets the tone for the entire production. In this plate, the water is black, only serving to make the novel ice stand out more, while the land in the distance is overshadowed by an ominous sky of dark clouds. The scene is melancholy, suggesting perhaps difficulties to come as the *Investigator*, alone under a threatening sky, weaves her way through ice floes seemingly impossible to navigate.

Figure 5.5 Samuel Gurney Cresswell, *Brown's Island, Coast of America*, August 1850. Watercolour, 18.2 × 25.5 cm. © Norfolk Record Office, Norwich.

A sketch exists by Cresswell entitled *Discovery of Barings Island September 6th 1850* (15.3 × 25.1 cm).[73] In this monochrome sketch, the composition is broadly the same as in the lithograph, with the shape of the ice in the foreground clearly transferred from sketch to print, although the ship is significantly larger in comparison to its environment. The accomplished lithographer William Simpson,[74] who became a professional artist shortly afterwards, has used his imagination to elaborate considerably on the details, creating something aesthetically pleasing out of Cresswell's rudimentary sketch. The reduction of the size of the ship, making the environment appear more overwhelming, is evident in all of Cresswell's lithographs where comparative drawings are available. A reviewer described the print thus: 'the little vessel is making its way through a vast field of broken ice, and beneath a sky so heavy with snow-clouds as to overwhelm the ship',[75] echoing the uneven battle of ship versus Arctic nature. In this combat between mismatched foes, it was then no fault of the British Navy if they lost, and it became a tribute to the 'stern courage and resolution' of those 'iron-hearted and iron-framed men'[76] if they triumphed.

It is unsurprising, then, that two plates (III and IV) show the *Investigator* keeling over dramatically in the ice. Both plates were published individually on 15 May 1854, before appearing as part of the set on 25 July 1854. The *Athenaeum* was no doubt referring to these plates when it commented: 'The few notes of the letterpress convey a forcible impression of the dangers and peculiar perils of Arctic voyaging; the heavy floes that crush a ship as if it were glass, or "nip" it into a shapeless heap of broken timbers.'[77] The third plate, *H.M.S. Investigator in the Pack. October 8th 1850* (Figure 5.6), described as a 'moonlit scene rendered with great power' in the *Art Journal*,[78] shows the position of the ship in the ice in October 1850, the first winter that the *Investigator* spent in the Arctic during McClure's voyage. The plate was lithographed by Edmund Walker and shows a scene that is almost Gothic in its rendering of the ice, with the ship's inner light glowing, reflected on the ice beneath a stormy sky; the half-concealed moon glimmers off the ice pack that itself displays striking contrasts of dark and light. The ice pack, eerie, glassy, indeed lurid, appears impossible to navigate either by ship or on foot, and no figures here leave the precarious safety of the vessel.

One of Cresswell's sketches, *Position of H.M.S. Investigator after Heavy Pressure 1852* (24.4 × 31.8 cm), looks similar in composition to *H.M.S. Investigator in the Pack* and may have been the inspiration, even though the print claims to show a scene from two years earlier.[79] Yet, in the

Figure 5.6 Samuel Gurney Cresswell, *H.M.S. Investigator in the Pack. October 8th 1850*, 1854. Lithograph, 44.3 × 61.2 cm. Courtesy of Toronto Public Library.

monochrome sketch, probably indicative of the only materials then available to Cresswell, two or three comparatively large figures walk on the ice, which appears more stable than in the lithograph. In the latter, the glow of light coming from the ship points to a spark of humanity that poignantly highlights the inhuman aspects of the ice pack.

The scene was also portrayed in one of Inglefield's lithographs, which he had published well before Cresswell's were printed and shortly after they both returned to Britain in October 1853 on the *Phoenix*. Four of the lithographs taken from Inglefield's watercolours were advertised as being in press at the end of October 1853.[80] These prints were promised to be 'of large and imposing proportions'.[81] The first of Inglefield's lithographs, *The Perilous Situation of H.M.S. Investigator, while Wintering in the Pack in 1850–51*, was 'taken from a Sketch by Lieut. [S.G.] Cresswell', and bears a resemblance both to Cresswell's sketch and to the lithograph *H.M.S. Investigator in the Pack.*[82] It is likely that both lithographs derive from the same sketch mentioned above, *Position of H.M.S. Investigator after Heavy Pressure 1852*. In both of the prints, the ship is considerably smaller than in the sketch and the icescapes are devoid of any outward human presence, excepting the ship (and a minuscule figure on board in Inglefield's version). The small size of the ship and the lack of people emphasise the immensity of nature in a way that Cresswell's sketch does not. Unlike in the sketch, the ice surrounding the ship appears completely unnavigable on foot, adding to the popular idea of utter imprisonment by ice. Furthermore, a type of 'against all odds' scenario is set up, during which the discovery of the Northwest Passage appears all the more impressive. The order of the prints in *Eight Sketches*, which otherwise follow a chronological sequence, suggests that the discovery (Plate VI) happened after the ship spent two winters in the Arctic, when in fact the expedition established the existence of the passage in October of the first year (1850).

The fourth lithograph in Cresswell's series (see Figure 0.2) is perhaps the most reproduced image of the nineteenth-century Arctic in the twentieth and twenty-first centuries; in particular, it is a popular choice to illustrate book covers on any aspect relating to the Franklin expedition and the searches.[83] It again shows the ship trapped in an immensity of ice the following year. Lithographed by Simpson, *Critical Position of H.M.S. Investigator on the North Coast of Baring Island, August 20th 1851* represents the vessel in an alarming position, appearing as though it might become crushed between masses of ice. Small figures are scattered on the ice that appears to crush the ship; some try to secure the ship to the ice while others may be placing gun powder in the ice in order to weaken it.[84] The reviewer in the *Art Journal* noted that 'the ship is imbedded between two enormous floes of ice as if they would crush her; this must have been a time of terrible anxiety to the navigators'.[85] We have no sketch from Cresswell dated 20 August 1851; neither is any surviving one similar to this lithograph. We do have access to several written accounts of events on this day, which describe how the *Investigator* was sailing between pack ice and land along the north shore of Banks Island. Whether the lithograph was actually representing events of 20 August is a matter of conjecture. Accounts in published narratives remind us that memory is fluid.[86] Johann Miertsching, the interpreter on the *Investigator*, makes no mention of the specific event represented in Plate IV, only noting on that day that 'she was anchored to a great stranded flow [*sic*] in the hope of pressing on to the east when the next friendly land-wind pushed the ice back from the shore'.[87] McClure writes in his journal that on 20 August they 'secured to the inshore side of a small but heavy piece of ice, grounded in twelve fathoms, seventy-four yards from the beach' as protection 'against the tremendous polar ice'. By evening, they were 'in a very critical position, by a large flow [*sic*] striking the piece we were fast to, and causing it to oscillate so considerably that a tongue, which happened to be under our bottom, lifted the vessel six feet'. The 'conflict continued for several minutes' when the large floe was 'rent into pieces' and the ship was 'driven nearer the beach'.[88] Alex Armstrong, the ship's surgeon, who used the lithograph as the frontispiece for his narrative, did note that on 20 August 'the ship's safety was suddenly threatened by a commotion in the ice . . . and now rendered our situation one of extreme danger'.[89]

Shortly after Cresswell arrived back in England, the *Illustrated London News* printed three engravings of his sketches with a summary of McClure's narrative of the voyage. They presented 'graphic and truthful pictures of adventures in the dreary, ice-bound regions of the Polar seas', as well as being 'illustrations of events described in Capt. McClure's despatches'.[90] One of the engravings, *Critical Position of 'The Investigator', at Ballast Beach, Baring Island* (Figure 5.7), bears a strong resemblance to the ever-popular fourth plate in Cresswell's series, *Critical Position of H.M.S. Investigator* (Figure 0.2).

The treatment of the sea ice differs in each, with the ice in the lithograph presenting a more vertical aspect reminiscent of icebergs in the eastern Arctic. However, both pictures show the ship in a very precarious position, and the existence of the engraving, printed within a few weeks of Cresswell's arrival in England, suggests that he did paint a scene of Ballast Beach, which may now have become lost or separated from the collection of paintings held by Norfolk Record Office. The *Illustrated London News* explained that this engraving showed where the *Investigator* was frozen up from 20 August to 11 September 1851.[91] It includes an extract from McClure explaining the situation (as above) when on 29 August 'a large floe, that must have caught the piece to which we were attached under one of its overhanging ledges, raised it perpendicularly thirty feet, presenting to all on board a most frightful aspect . . . This suspense was but for a few minutes, as the floe rent.'[92]

However, a sketch by Cresswell does exist of the *Investigator* some weeks later in a position of difficulty, entitled *Position of H.M.S. Investigator, Sept 19, 1851* (17 × 24.5 cm).[93] Here, the ship is being forced at an angle by the sea ice, and men are anchoring her to a floe. The ice is in no way as high as it was represented in either the lithograph or the engraving. In fact, none of Cresswell's paintings from the voyage shows

Figure 5.7 Samuel Gurney Cresswell, *Critical Position of 'The Investigator', at Ballast Beach, Baring Island*, 29 October 1853. Engraving, 11 × 15.5 cm. © Illustrated London News / Mary Evans Picture Library.

ice as high and vertiginous as in the two prints. Cresswell's sketch shows the figures, who anchor the ship to a grounded floe, in the foreground of the scene, making them much larger than they appear in the prints when they retreat to the background of the scene, appearing tiny and helpless, although they too are engaged in securing the ship to the ice. The figures in the sketch appear to be more in control of the situation.

Armstrong recounts that this day was filled with apprehension, when masses of ice lifted the ship fourteen inches out of the water while she was anchored to a large floe. The continual assault of the ice 'presented a prospect of peril' throughout the day.[94] It is interesting too that the sketch is simply entitled *Position of H.M.S. Investigator, Sept 19, 1851*, whereas the lithograph and engraving captions have prefixed the word 'Position' with 'Critical', emphasising the danger for readers. Inglefield's lithograph too, made use of the descriptor 'Perilous' in the title, thereby also clarifying the strong possibility of the wreck of the ship for the viewer.

Reviewers were concerned with accuracy and truth; for the most part, the lithographs carried an almost irrefutable stamp of authority. Cresswell's substantial archive indicates that visual records could be transformed through lithography, with Simpson and Walker having to fill in detail where none existed. The selection of his work chosen for printing was judicious, displaying themes of discovery and danger over any other aspects of the Arctic expedition. Material showing sailors who might appear to be lounging about, instead of gallantly striving to rescue the lost expedition, was not suitable for publication. Furthermore, Cresswell's departure for the Baltic shows that these amateur naval artists were not necessarily able to oversee the production of lithographs from their work. Lithography artists were excellent artists in their own right; shortly after working on Cresswell's prints, Simpson became a pioneer war artist in the Baltic, and thereafter Queen Victoria regularly commissioned his work.[95]

Walter William May's *Series of Fourteen Sketches* (1855)

When Walter May returned to the Arctic as lieutenant on the *Assistance* as part of the Belcher expedition of 1852 to 1854, it was his second expedition in search of Franklin.[96] He illustrated the 'Queen's Illuminated Magazine and North Cornwall Gazette'[97] (as discussed in Chapter 2) and also undertook other drawing as a pastime during the voyage. Some of his drawings from the Belcher expedition were subsequently published as a folio of lithographs on 1 May 1855, showing 'fourteen most interesting and carefully-finished sketches'.[98] May 'intended to illustrate a few of the principal events and features' of the expedition, including 'some in a picturesque point of view, and others to illustrate "Arctic Travelling", which though ably explained to the public in other works, has not yet been produced as one of the principal objects in a series of Sketches'.[99] In his use of the term 'picturesque', he here signals the hope that the prints will be aesthetically pleasing, appeal to a broad audience, and not just tell a narrative of the expedition.

A close study of May's drawings, and the lithographs derived from them, reveals the ways in which the Arctic presented to the public subtly changed once it arrived in Britain. By examining the lithographs, their written descriptions by May, and his drawings and paintings in the Arctic, we see the Arctic change into a darker, more sublime space, particularly in the prints that illustrate 'Arctic Travelling', than the one initially represented by May in other formats. While the prints are clearly different from the drawings, the form of their differences changes the depiction, and meaning, of the Arctic in several ways. More specifically, the printed and published Arctic became more physically difficult and inhuman, more imperial and exotic, than the Arctic depicted in the sketches and watercolours.

Although the lithographs show a struggle of heroic failure, other sources indicate that May was enthusiastically immersed in, and at ease with, Arctic life. His drawings, his own journal, and his appearances in other texts reveal much of a personal connection to the Arctic.[100] All the work discussed below is drawn from May's second Arctic voyage, when he acted as lieutenant on the *Assistance* under the notorious Edward Belcher from 1852 to 1854. May's own drawings and watercolours, from which the lithographs are derived, show a far more cheerful Arctic than the one we see in print.

The Belcher expedition had many fine, clear days in its first winter quarters, despite being one of the farthest north of all the search expeditions, and May made the most of this, even drawing outside in winter, commenting that 'the weather being so fine with no wind we do not feel the cold'.[101] This fine weather made it possible for the expedition members to spend time outside, exercising, hunting, and recording meteorological phenomena in winter. It also facilitated enjoyment and appreciation of the environment. Nowhere in May's personal journal does one get a sense of imprisonment in the Arctic winter or the alienation and loss of self that is associated with the negative sublime. However, the lithographs that were published after his voyage would be appreciated, one reviewer felt, as a record of 'undaunted bravery and heroic perseverance under difficulties of no ordinary nature'.[102] May himself eschewed hyperbole and the language of the sublime in his text, writing in an unadorned style without exaggeration: 'Among the plates I shall now try to give a truthful description of the mode we pursued in Arctic Travelling.'[103] He also fails to use the word 'exploration', instead choosing the more ordinary term 'travelling'.

May's folio of lithographs, *A Series of Fourteen Sketches*, was published by Day and Son on 1 May 1855 at a cost of twenty-one shillings (or one pound and one shilling).[104] The folio includes a list of subscribers of 240 names.[105] Some subscribers ordered more than one copy, leading to a total production (as per the list) of 259 copies. Almost one-quarter of the subscribers were women, indicating a significant female interest in viewing and owning representations of Arctic exploration at the time. The following list details the titles of the plates in the folio and shows their focus on ships and sledges.

A Series of Fourteen Sketches: List of Plates by Walter William May/ Day and Son

I.— The Arctic Squadron in Lievly Harbour, Island of Disco, West Coast of Greenland.
II.— Loss of the *McLellan*.
III.— H.M.S. *Assistance* and *Pioneer* in Winter Quarters – Returning Daylight.
IV.— H.M.S. *Assistance*, in Tow of the *Pioneer* (Captain Sherard Osborn), Passing John Barrow Mount, North of Wellington Channel, 1853.
V.— H.M.S. *Assistance* and *Pioneer* fast to the floe, off Cape Majendie, Wellington Channel, 1853.
VI.— Perilous Position of H.M.S. *Assistance* and *Pioneer*, on the Evening of the 12th of October, 1853.—Disaster Bay.
VII.— H.M.S. *Assistance* and *Pioneer* Breaking out of Winter Quarters, 1854.
VIII.–IX.— Division of Sledges Finding and Cutting a Road through Heavy Hummocks, in the Queen's Channel.
X.— Division of Sledges Passing Cape Lady Franklin; Extraordinary Masses of Ice Pressed against the North Shore of Bathurst Land.
XI.— Sledges in a Fresh Fair Wind, Going over Hummocky Ice.
XII.— Encamping for the Night.
XIII.— Sledge Party Returning through Water during the Month of July.
XIV.— Relics Brought by Dr. Rae.

A number of unique drawings exist that can be shown to be prototypes of the published prints. Several of these are in the National Maritime Museum, Greenwich, but similar themes can be found in illustrations by May in the 'Queen's Illuminated Magazine', and additional drawings by him exist that could also have been prototypes for the lithographs. A set of drawings that provides the prototypes for three of the six lithographs concerning sledge travel was likely to have been made after the voyage.[106] In the first instance, they are done on paper imprinted with Ackermann's name. The paper is clean and undamaged, unlike the material made in the Arctic that tends to be worn, folded, or marked, reflecting its contextual origins. Furthermore, pictures done in the Arctic tend to be on varying sizes of paper, their makers using whatever was available to them at the time. These three drawings are done on approximately the same size paper (27 × 35 cm).

In addition, the way in which May writes about the plates in the published folio often indicates that some drawings were created after the event, probably in England. These 'Arctic Travelling' scenes may have been included at the suggestion of another agent. Their use of the same paper, stylistic differences, and medium marks them out as separate from the rest of May's work. In plates I to VII, May is explicit in stating the pictures' origins. For example, in *IV.—H.M.S. Assistance, in Tow of the Pioneer*, May tells us that the 'sketch was taken on one of those beautiful calm evenings'.[107] Indeed, a worn and well-travelled watercolour exists in the National Maritime Museum, Greenwich, that shows a similarity to the lithograph.[108]

In the drawings and lithographs, the actions, facial expressions, and even the number of figures convey different versions of Arctic exploration. The presence of human figures in many of May's sketches and lithographs provides a barometer of change as representations went through processes of transformation. In *XIII.—Sledge Party Returning through Water in the Month of July*, certain aspects of the composition change between painting and print.[109] May tells the reader in his account of the drawing: 'The Sketch in this Plate is intended to represent Captain Richards' party returning across Byam Martin Channel, in the month of July. They are supposed to be bringing the *Pioneer*'s ice-boat from Melville Island, two hundred miles from the ship, where she had been placed by one of his depot sledges in the spring.'[110]

May is clear here that in fact this is an imaginative composition, based on his knowledge of travel and the ice. The plate is 'intended' to represent a scene and the figures are 'supposed' to be doing a certain activity. Despite this, it was felt necessary by someone to heighten May's composition to affect a more sublime Arctic. Both the watercolour by May and the subsequent lithograph show the unpleasant work of hauling a sledge through the meltwater that forms on sea ice during the summer period. Although the lithograph is far more dramatic in many ways, from the angle of the boat to the setting sun,[111] the focus here is on the first figure pulling the sledge in both pictures, and it is obvious that the printed version shows a figure engaged in a far greater struggle, with his back bent at such an acute angle.

This would seem to suggest a great war is being waged against the ice in order to find Franklin and that everything humanly possible is being done in the Arctic search. This tendency to exaggerate the angles at which figures strain to pull a heavy load over the ice is also seen in other lithographs of the set, for example in *II.—Loss of the McLellan* and *IX.—Division of Sledges Finding and Cutting a Road through Heavy Hummocks*. The figures show a valiant struggle wholly compatible with the idea of 'gallant' crews striving to conquer the Arctic.

But it is not just the angles of the figures that change, their faces can also be transformed in the printed material. The lithographs do not show facial expressions in the same way as source drawings or watercolours. An examination of the watercolour *Loss of the McLellan* (1852) and the lithograph *II.—Loss of the McLellan* (1855) also reveals subtle yet important differences, although at first glance the two pictures appear to be quite similar.[112] The ships in both remain in much the same places and at the same angles; the boats in the foreground too seem to mimic each other; the horizon line is unchanged.

Once again it is evident by noticing the angles of the figures that those in the watercolour (31.6 × 46.7 cm) are not struggling to the same degree as the ones in the published lithograph. This includes the smaller figures in the background. However, looking more closely at the watercolour (Figure 5.8, top), we can see smiling expressions on the two men who face the artist as they heave the boat over the ice; one of them, distinctly comical with a red nose and what seems to be an eyepatch, looks right at the artist. These faces show a very human and personal Arctic, whereas the lithograph, with its hidden faces, projects a depersonalised context, making it less of a place, with its own community, and more of a space with which to battle. Indeed, it is rare to find any facial expression in published individual prints of the Arctic, although on-board newspapers are full of smiling faces that can be seen in the *Illustrated Arctic News* and in the 'Queen's Illuminated Magazine'.

While the loss of a ship in the ice may sound catastrophic, in this case the crew, who were surrounded by relatively stable ice in Baffin Bay and a large amount of whaling ships as well as the search expedition, were easily rescued. The *McLellan* was in a far better position than trading and passenger vessels much further south in the summer of 1850 that ran into icebergs in the Atlantic. For example, Irish newspapers reported the destruction of fourteen vessels and the loss of over one hundred lives, many of whom were on an unidentified vessel (reported to have been from Derry) sailing to Québec that sank amidst the ice over several days. Although ice was generally watched out for in the Atlantic in April and May, owing to the spring break-up of the Arctic seas, the year of 1850 was exceptional and the 'floating fields of icebergs' were 'immense'. 'Subsequently a great many bodies were seen intermingled with the ice,

Figure 5.8 Detail from Walter William May, *Loss of the McLellan*, 1852. Watercolour; pen and ink (top). Detail from Walter William May, *Cape Lady Franklin*, 1852/4. Graphite (bottom). © National Maritime Museum, Greenwich, London.

together with some portion of the cargo.'[113] Such a tragedy shows that any seafaring, not just Arctic exploration, was a risky business in the nineteenth century (as indeed it can still be today). In sharp contrast, May's textual description of this plate in his folio relates that 'the Sketch was taken when the men were deserting the wreck; some were launching their boats to a place of safety. Boxes, beds, casks, clothes of every description, were scattered about the floe. It formed a most animating scene, and made a capital subject for a picture.'[114] Although May does not register any sense of danger in the accompanying text, which makes more sense when viewed in conjunction with the happier figures in the watercolour, the plate shows a far more threatening Arctic epitomised by the labour and struggle of the men. McDougall, too, noted the implausibly relaxed attitudes and the peculiarity of this scene: 'It was novel, but interesting, to gaze on so many vessels in a state of utter helplessness, careening and fouling each other in every possible direction, whilst their crews, standing beside their boats and clothes on the ice, smoked their pipes like perfect philosophers.'[115] The incident also appears pictorially on the front page of the January 1853 issue of the 'Queen's Illuminated Magazine', entitled *A Nip in Melville Bay.*[116] In the magazine, the account of the incident beneath refers to the picture as being 'the next in our picture gallery', which somewhat trivialises the incident. However, Sherard Osborn reveals that 'Piling the agony! might we feel be pardonable in such a case, but that has so often been already done, in all connected with Polar Voyages, that we shall endeavour to give a ... matter of fact account.'[117] In May's pen and ink picture in the on-board periodical, the individual men with their possessions are fore-grounded, and the pulling of the boat is relegated to the background. In fact, the entire Belcher expedition, consisting of five ships, witnessed the loss of the American whaling ship, the *McLellan*, to the ice pack off the west coast of Greenland, and the incident was recorded visually by several expedition members. The shipwreck took place over a period of ten days in July 1852, and May's entries in his journal over the extended period are factual and calm, showing no trace of astonishment or awe associated with the sublime. In fact, with the ships docked in the ice, he spent time trying to prepare paper for the calotype, an early photographic process.[118]

In the same way that the panorama peopled its Arctic with frenzied activity to imply the large number of men engaged in outdoor 'manly' pursuits, some of May's lithographs, too, increase the number of human figures. The sketch for *IX.—Division of Sledges Finding and Cutting a Road* attributed to May is one of the three sketches mentioned above. It has the appearance of being produced directly for the purpose of preparing a lithograph and is part of a set of similar drawings that are all done in the same medium, on the same paper, which is embossed with the name of the lithography company 'Ackermann & Co.' It is clear that five extra figures were added into the sketch by using a hard pencil as opposed to the chalk pastel used for the rest of the scene. In particular, two of the added figures stand on a height gazing out over a vast landscape, suggesting an unexplored space beyond and a mastery of the landscape that could be possible through labour and struggle.

When May's drawings were reinterpreted as lithographs, it was not only human forms and faces that changed; the landscape was also altered and intensified. It was through the landscape that the sublime associations of the Arctic could be conveyed and popularised most effectively by drawing on Romantic and Gothic traditions. By comparing another preparatory drawing attributed to May, *Division of Sledges Passing Cape Lady Franklin*, with the corresponding print from 1855,[119] we observe how the printed landscape has sharper peaks and high contrasts, reminiscent of the treatment of Browne's landscapes in the panorama. This published version shows a landscape that is altogether darker and more threatening, with its backdrop of gathering storm clouds. A closer look at the drawing also reveals the smiles on all three visible faces in the drawing (Figure 5.8, bottom), which recall the smiling faces in the watercolour *Loss of the McLellan*. The figures in the drawing are, without exception, turned towards their partners, reminding us of the camaraderie, humour, and enjoyment experienced in the Arctic environment. Those in the lithograph, however, are more engaged with the struggle of their endeavour, presenting a darker face and anonymity, a loss of self in the Arctic.

It was unsurprising then that the *Morning Chronicle*'s reviewer of the lithographs could only see the 'intense misery, the excessive labour, and the constant perils which the naval heroes who have adventured into those frozen latitudes have been compelled to undergo'.[120] The lithographs correspond with what was being written in the media about the expeditions, and the prints match the image of the search for Franklin as a trial of difficulty and self-sacrifice in a way that smiling faces in sketches, paintings, and drawings did not. Like the labourers in early nineteenth-century English landscape paintings, these industrious men have become reassuring representatives of the moral character of their country.[121] In fact, the lithographs suggest the nobility of the moral sublime even more forcefully than it had appeared in the panorama *Summer and Winter Views of the Polar Regions*.

Given that May's folio intended to represent Arctic modes of travelling, it is noticeable that no sledge dogs are included in the plates. However, the Belcher expedition regularly used dogs for pulling the sledges, particularly when travelling between the *Assistance* and the *Resolute*, and several men became adept dog drivers.[122] During the second winter, the route from the *Assistance* near Cape Osborn to the *North Star* at Beechey Island, a distance of around a hundred kilometres, was travelled repeatedly by dog sledges that were used as 'express couriers'.[123] Their absence in the lithographs is striking, and the reasons for not including them are puzzling. Perhaps whoever was influencing the publication of the lithographs did not think that the dogs portrayed the right message. The dogs do appear on two occasions in the 'Queen's Illuminated Magazine', pulling a sledge driven by a man in the small pen-and-ink *Snow Carriers*,[124] and four dogs take centre stage in a watercolour (16 × 23.5 cm) attacking a bear that May shot during the first winter.[125] This exclusion of the dogs from the lithographs (none of the lithograph folios shows them) has left the image of man struggling in the snow, battling against nature, valiantly pulling a sledge, as the transport that is associated with British polar exploration. William Barr has calculated that dog sledges as opposed to man-hauled sledges undertook some 28 percent of the overall distance travelled by sledge during the British Franklin search expeditions.[126]

The notion of the moral superiority of man-hauling and the struggle that we see in May's lithographs, but not to the same extent in his drawings, was adopted by Robert Falcon Scott and re-enacted in Antarctica in the early twentieth century.[127] As Lisa Bloom argues, Scott was concerned with 'constructing an image of a noble struggle', and dogs 'would compromise this heroic image'.[128] These lithographs and other prints like them were a key way of reinforcing the polar regions as a 'theatre of the tragic-heroic defeat of hubristic aspiration, a figuration which is central to the remediation of the polar regions in both the *Ancyent Marinere* and *Frankenstein*'.[129] In many of May's lithographs, the viewer is presented with a degree of difficulty and physical pain indicating a Burkean sublime that is absent in the preliminary drawings. The smiling expressions of the drawings negate the sublime; solitude and silence for example, were virtually impossible on those expeditions, where a large homosocial society nestled amidst the vast skies, the smiles reminding the viewer that the men knew each other well through shared experience. The largest expeditions consisted of hundreds of men, and some smaller expeditions wintered near Indigenous communities with whom they socialised. The drawings were not fully compatible with the 'romantic glamour ... that attaches to the idea of taking risks and pains in one's travelling'.[130] In essence, the suffering apparent in May's lithographs on Arctic Travelling can be seen as performative suffering within a sublime landscape. The visual absence of the dogs, or any reference to them in the written account of the plates, signals a desire on the part of whoever was influencing the publication of the lithographs to display a more perilous and difficult travelling experience, one of the '*markers* of authenticity, and flowing from that authenticity an authority and cultural capital'.[131] The lithographs present the Arctic, foreshadowing Scott's heroic suffering in Antarctica, as the 'theatre of heroic defeat'.[132]

Conclusion

With their comments on the 'appearance of inaccuracy' in Cresswell's lithographs, the reviewer in the *Athenaeum* had, perhaps unwittingly, hinted at the difficulty that lithographers faced when transforming an amateur artist's sketches of the Arctic into an appealing and convincing product. It is likely that this required extensive input from the lithographic artists' imagination and memories, including, perhaps, a familiarity with the numerous Arctic exhibitions in London that had been opened to the public since 1850.

The display of expensive lithographs in printsellers' windows and inside their premises allowed a far greater number and variety of people to view them, indicating that the influence of the Arctic lithographs was greater than has previously been assumed. The high cost did not preclude their impact on the people who could not afford to have them on their drawing-room tables. The format of these productions, '*fac similied*' from officers' drawings suggested a commitment to both 'truth' and 'accuracy' and the association of lithography with 'direct duplication'[133] reinforced that impression. Viewers must have been aware of the tendency of exhibitions such as panoramas and dissolving views to exaggerate or be based on an imaginary Arctic, but the lithographs were trusted, for the most part, as being wholly factual. However, the unease surrounding the truth-value of pictures, which extended to such lithographs even though they were generally thought of as being reliable, suggests a distrust of visual material as a factual source by mid-century.

While Browne's folio *Ten Coloured Views* (1850) betrays a strong scientific interest through composition, subject matter, and titles, it also conveys a romanticism that connects with nature. However, despite the topographically impressive cliffs of the region, the lithographer still felt the necessity to heighten any sublime effects, particularly to enhance the darker aspects of the search. Later lithographs of Inglefield, Cresswell, and May increasingly imply that a battle is being waged against the capricious Arctic nature in an effort to find Franklin. They show an interest in displaying British masculinity, ships in peril, and the labour of sledge travel suggesting the moral sublime.

Ultimately, these prints must be viewed as artefacts that combine officers' sketches, lithographic artists' imaginations, and the interests of the Admiralty and of the publishers. Arresting images like Cresswell's famous *Critical Position of H.M.S. Investigator*, which are prevalent in secondary texts today, were not the products of individuals alone.

CHAPTER 6

Conclusion
Resonances

If the Franklin search could be distilled down into one representation in popular culture, it is this: *Critical Position of H.M.S. Investigator* (44.3 × 61.2 cm) (Figure 0.2), created by the lithographer William Simpson and based on the work of Samuel Gurney Cresswell. The iconic image of the ship keeling to one side, about to be crushed by mountains of ice, the tiny faceless, depersonalised figures lost under a stormy sky, is frequently reproduced. It is no coincidence that, of the many hundreds of pictures produced from the Franklin searches, this one shows intense danger, heightened sublimity, and dark and light contrasts – the visual markers we have been led to expect on every heroic historical Arctic expedition. As Lisa Bloom has commented, 'the difficulty of life in desolate and freezing regions provided the ideal mythic site where men could show themselves as heroes capable of superhuman feats'.[1] We have inherited a genealogy of commercial visual material from the nineteenth century that continues to multiply digitally. Images of ships imprisoned by the Arctic ice, darkness, and ferocious beasts resonate long after the details of the words have been forgotten.

By contrast, the many unpublished visual records stored in archives remain a forgotten part of the narrative. In part this is due to their presence in other documents, such as personal journals, and, thus, their contents are not individually catalogued. *Visual Culture and Arctic Exploration* has established the visual content and extent of neglected archival records from ships and examined these in light of more public and well-known representations of the Arctic. By combining practices and ideas from the disciplines of literature, visual culture, the history of art, and historical geography, I have analysed pictures and texts to uncover the nature of the on-board and metropolitan representation of the Arctic. This book has paid close attention to media type, contexts, and reception, moving from the visual culture aboard the ship to that of the metropole, noting that representations can change dramatically between the two venues. The sharp focus on visual material counteracts the tendencies of cultural studies to focus on purely literary representations of the Arctic. By examining a wide range of media produced within a relatively short time span, this has contributed to knowledge beyond the broad surveys of visual culture with their emphasis on public representations.

Visual culture was deeply embedded in shipboard life, and the practice of drawing extended to all ranks on board. Although we would expect naval officers to produce topographical and scientific drawings, many drew for more personal reasons: to create a souvenir or keepsake, or to entertain their fellow expedition members, for example. Such pictures often have affective resonances, and even pictures done for more scientific purposes may show subjective responses. Illustrated periodicals produced on board mined the expedition members' situation to create a specifically Arctic humour, one that poked fun at their own interaction with the Arctic environment and overturned the sublime. These periodicals were not just created to pass the time during the long winter 'imprisonment' in the ice; they were also conceived of as mementos, to remind expedition members of 'old Arctic days',[2] suggesting an emotional attachment to their time in the Arctic. By examining on-board pictures, in conjunction with their associated texts, a far more complex visual matrix is revealed than that which entertained audiences in the metropole. This matrix incorporates traits such as familiarity, domesticity, humour, and emotional attachment, characteristics that were not generally associated with Arctic exploration.

Unlike the material produced on board, much of the metropolitan visual culture portraying the Arctic leaned heavily on imagination and on the lure of the sublime. Despite assertions that prints and panoramas used officers' on-the-spot sketches as their source, such pictures were altered as they were transformed for publication and exhibition. The result was a heavily coded Arctic, one that emphasised danger, incessant labour, masculinity, and the ever-present threat of shipwreck amidst towering icebergs. As the search wore on, the activities of expedition members were increasingly represented as a battle with nature, one in which icy wastes became an alibi for failure.

Just as we prefer them now, stories of adventure, danger, and death predominated in the nineteenth-century media. Positive images, surviving amply in the primary record, were certainly less well represented by the media. This was noticed at the time as being misleading by expedition members themselves and the transformation of the Arctic experience, in words and pictures, was apparent to those who participated in the expeditions. Edward Belcher commented on this in his narrative *The Last of the*

Arctic Voyages (1855), when he described the loss of the hired transport ship, the *Breadalbane*, in the ice in 1854. This event, he surmised, 'will doubtless be magnified into something grand and sublime, perhaps got up for amusement at one of the minor theatres, and frighten the loving wives of some of our married men'.[3] Even Sherard Osborn, whose narrative *Stray Leaves from an Arctic Journal* (1852) gives a largely positive view of the Arctic, confessed: 'I am free to acknowledge, I have "piled the agony" to make my work sell.'[4] Osborn here acknowledges the appeal of 'agony' and the attraction of the human mind to representations of pain. Isaac Israel Hayes, who published a narrative of a harrowing boat journey on Elisha Kent Kane's disastrous expedition, noted that the polar regions were associated with 'the wildest fancies and the most repulsive conclusions'. This was, he felt, due in part to the fact that 'the expeditions which have particularly attracted the general notice of the civilised world, have been the disastrous ones'. He described the published reports of the Arctic regions thus: 'Vast seas covered with masses of ice rushing to and fro, threatening to crush the most skilful navigator – towering bergs ready to overwhelm him – dangerous land journeys – cold, piercing to the very sources of life – savage beasts, and scarcely less savage men – isolation, disease, famine, and slow death.'[5]

The close analysis of visual records from the Franklin searches has revealed the complexity behind nineteenth-century prints and exhibitions. As pictures transform, so too do their associated words. As skies darken, ships' positions become 'perilous'. Words and pictures work together to tell a convincing story in a commercial marketplace. These transformations from private to public remind us that nineteenth-century publications cannot be seen solely as representing the perceptions of 'explorers'. By closely reading contemporary reviews, the accessibility and power of an Arctic panorama has been revealed, and my research shows that expensive items like large colour lithographs could be seen, if not necessarily purchased, by all social classes. Reviews also betray a concern with 'truth' in visual representation, suggesting that there was an underlying distrust of popular visual culture as a source of information. Victorian culture was an integral part of shipboard life, exemplified in theatricality, lavishly illustrated periodicals, and the practice of drawing and writing that was not confined to the higher officer classes. Indeed, the Arctic search ships can be viewed as microcosms of Victorian culture. By examining the archive, I have demonstrated that drawing, as much as writing, was a key part of expedition members' response to the environment. Its practice was as important for personal reasons, for emotional ones, as it was for scientific ones or to add to knowledge. The impact of colour imagery, still novel in the mid-nineteenth century, aboard a ship overwintering in the Arctic is significant in recognising the centrality of the visual to the expeditions.

The image of the nineteenth-century Arctic – an eternal space of winter associated with man's 'against-all-odds' battle with nature – is complicated by the evidence. In the archive, that Arctic is often represented as a familiar, local, social, and humorous place. Expedition members' desire to create 'souvenirs' and the expectation of future nostalgia are evident in the visual and written archive. As Carolyn Strange argues,[6] polar exploration, in this case the refuge of the Arctic winter, could inspire genuine cheerfulness and humour. The winter, thought of negatively in the media as an 'icy prison', was often referred to as a 'sojourn' by expedition members, and the darkness provided a time to rest before the sledge journeys of spring. Domesticity and 'feminine' activities, which we would not generally associate with polar exploration, were necessitated by the experience of wintering in the Arctic. This mode of hibernation is well represented through the intimate, personal, and humorous depictions of winter life shown to us in the illustrations and text of on-board periodicals. For the public back in the metropole, however, the Arctic winter was more likely to be associated with horror, pain, and difficulty.

The study of visual culture and travel literature benefits from an interdisciplinary approach, one that attends to human and physical geographical contexts, employs art historical analyses, and incorporates close readings of image and text together. By working across disciplines, other aspects of the visual and cultural histories of Arctic exploration have been brought to light. In-depth archival work reveals the importance of tracing, where possible, the archival sources for published versions; attending to geography shows the important regional differences in the Arctic that affected its representation; examining pictures and texts together can reveal disjunctions that signal more complicated histories.

Richard C. Powell has commented on the absence of twentieth-century geographies of the North. Furthermore, the Arctic seems to lend itself to a 'depressingly simplistic narrative that dominates public understandings of the circumpolar north'.[7] The modern polar imaginary is dominated by Antarctica. Indeed, in Ireland the historical narrative revolves around the heroism of two main actors: Tom Crean and Ernest Shackleton. In particular, the study of the visual culture of Arctic exploration has been neglected, which has implications for how we view the polar regions today. Now, as in the nineteenth century, the modern perception of the circumpolar North, if any, is of a region largely empty, eternally covered in ice and snow. The legacy of the nineteenth-century imagery of perpetually snow-clad regions still reverberates in the modern imagination. This perception of the North as an eternally cold and icy wasteland makes it easy for people to disregard the fact of global warming and to continue with their everyday lives. With climate change in the Arctic more severe than elsewhere, many implications are evident: human impacts are being felt in the region, and, at the same time, the opportunities for economic exploitation provided by ice-melt highlight the contested nature of Arctic space and increasingly threaten the traditions of Indigenous peoples.[8] The words written in the booklet that accompanied the panorama *Summer and Winter Views of the Polar Regions* (1850) now seem strangely prophetic: 'for all purposes of commerce or traffic, the Arctic regions must remain a sealed book to the human race, until some vast change shall occur in the temperature and in the meteorological arrangement of the globe.'[9] One hundred and seventy years later, that 'vast change' is upon us.

The discovery of Franklin's ships highlights the problematic use of Franklin and the search expeditions to support claims for Canadian sovereignty over the waters of the archipelago, as the claim of Canadian sovereignty over the Northwest Passage is contested by other countries such as the US, which views the passage as an international strait. If Canada can demonstrate that the Northwest Passage constitutes 'historic internal waters' citing centuries of British exploration as evidence of use, the country will have a stronger claim for sovereignty over the archipelago.[10] However, as Byers has stressed, 'the strongest element in Canada's claim is the historic occupation by the Inuit, who have hunted, fished, travelled and lived on the Northwest Passage for millennia'.[11] Adriana Craciun argues that the locating of the *Erebus*, after six search seasons, was presented as the 'lynchpin in Canada's historic sovereignty of the Arctic archipelago and of the Northwest Passage'.[12] In addition, the sponsorship of Shell, which owns a large number of oil and gas leases in the North American Arctic, in the search seeks to connect their energy exploration and exploitation of the Arctic to the heroism associated with Victorian Arctic exploration.[13]

The emphasis on the 'man versus nature' trope, heroic failure, and suffering in the published material, particularly in the visual material, has become the standard mode of viewing the Arctic. Our present-day relationship with the Arctic, and with its history, is still dominated by these published pictures and texts, while the on-board histories – local, intimate, and domestic – have remained largely concealed. This book has uncovered and examined records of the Franklin search expeditions that previous critical scholarship has largely overlooked, thereby revealing a version of the Arctic search that is far more complex than a story of man's battle against nature.

North Star Mount (Figure 1.8) (14 × 22.5 cm), a small watercolour (likely faded with time) in the National Library of Australia, serves to remind us that the nineteenth-century Arctic was more than an icy wasteland, a theatre for the moral sublime.[14] The torn, unsigned, and unassuming picture that shows an Arctic summer is only one fragment of a large visual archive that tells the story of the Franklin search in a very different way, an archive that is fragile, neglected, and scattered around the globe. Perhaps the green hills, calm waters, and fresh blue skies depicted here off northwest Greenland will explain the reaction of a midshipman from Cobh, Co. Cork, aboard the *Herald* in Panama on hearing news of their orders to participate in the search for Franklin: 'We were all, of course, in high spirits at so unexpected and welcome a change. As for myself, I was delighted; nothing could be more charming and novel than a cruise to the Arctic regions.'[15]

Notes

Introduction

1. The panorama was a huge painting comprised of canvases enclosed in a circular building known as a rotunda, where the spectator contemplated the view from a raised platform in the centre of a circle. The word was coined by Irishman Robert Barker to describe his circular painting of Edinburgh and the surrounding countryside in 1789. Individual panoramas at Leicester Square often exhibited for a year.
2. 'Panorama of the Arctic Regions', *Observer*, 10 February 1850.
3. On 19 May 1845, the *Terror* and the *Erebus* sailed from the Thames towards the Canadian Arctic archipelago. Europeans last saw the expedition at the end of July in Melville Bay (off the west coast of Greenland). All of the 129 men on board died during the expedition. Neville W. Poulsom, *British Polar Exploration and Research: A Historical and Medallic Record with Biographies 1818–1999* (London: Savannah Publications, 2000), 59. Russell Potter notes that, in the last twenty-five years, the Franklin story has inspired 'more than two dozen novels, three documentaries, various plays, poems, and radio dramas, an Australian musical, and a German opera'. *Finding Franklin: The Untold Story of a 165-Year Search* (Montreal and Kingston: McGill-Queen's University Press, 2016), p. xvi.
4. The number of expeditions here is taken from W. Gillies Ross, who provides an excellent quantitative analysis of the search efforts. 'The Type and Number of Expeditions in the Franklin Search 1847–59', *Arctic* 55, no. 1 (March 2002): 57–69.
5. Historian of Science Nanna Katrine Luders Kaalünd draws attention to the fact that nineteenth-century British Arctic expeditions were 'inherently international projects' that included a reliance on Indigenous peoples, foreign national expedition members, and the knowledge acquired by previous expeditions, regardless of nationality. 'What Happened to John Franklin? Danish and British Perspectives from Francis McClintock's Arctic Expedition, 1857–59', *Journal of Victorian Culture* 25, no. 2 (2020): 300–14 at p. 301.

6. Chauncey C. Loomis, 'The Arctic Sublime', in *Nature and the Victorian Imagination*, ed. U. C. Knoepflmacher and G. B. Tennyson (Berkeley and Los Angeles: University of California Press, 1977), 95–112 at p. 96.
7. William Jerden (ed.), 'Panorama of the Polar Regions', *The Literary Gazette: A Weekly Journal of Literature, Science, and the Fine Arts*, 16 February 1850.
8. Richard Cyriax, *Sir John Franklin's Last Arctic Expedition: The Franklin Expedition. A Chapter in the History of the Royal Navy* (London: Methuen, 1939) and Ann Savours, *The Search for the North West Passage* (London: Chatham, 1999) are both excellent on the topic.
9. Savours, *North West Passage*, 56.
10. Andrew Lambert, *Franklin: Tragic Hero of Polar Navigation* (London: Faber & Faber, 2009), 151; 166.
11. Savours, *North West Passage*, 186–9.
12. The Inuit homeland of Canada consists of four regions together called Inuit Nunangat. Nunavut and the Inuvialuit Settlement Region include the entire Canadian Arctic archipelago, while Nunavik covers northern Québec, and the Inuit region of Labrador is called Nunatsiavut. Inuit Tapiriit Kanatami, 'Inuit Regions of Canada' (Ottawa, 2019). www.itk.ca/about-canadian-inuit/
13. For example, monographs include Francis Spufford, *I May Be Some Time: Ice and the English Imagination* (London: Faber & Faber, 1996); Eric G. Wilson, *The Spiritual History of Ice: Romanticism, Science, and the Imagination* (Basingstoke: Palgrave Macmillan, 2003); Peter Davidson, *The Idea of North* (London: Reaktion, 2005); Jen Hill, *White Horizon: The Arctic in the Nineteenth-Century British Imagination* (Albany: State University of New York, 2008); Hester Blum, *The News at the Ends of the Earth: The Print Culture of Polar Exploration* (Durham, NC: Duke University Press, 2019). More recently, the volume edited by Heidi Hansson, Maria Lindgren Leavenworth, and Anka Ryall, *The Arctic in Literature for Children and Young Adults* (London: Routledge, 2020), makes a valuable contribution to the literature.
14. The work of art historian Ingeborg Høvik is a notable exception, including 'Arctic Images 1818–1859' (PhD thesis, University of Edinburgh, 2013); 'Framing the Arctic: Reconsidering Roald Amundsen's Gjøa Expedition Imagery', *Nordlit* 25 (2015): 137–60; 'Reproducing the Indigenous: John Møller's Studio Portraits of Greenlanders in Context', *Acta Borealia* 33, no. 2 (2016): 166–88; 'Girlhood in the Arctic: Word-Image Relations in R. M. Ballantyne's Canadian Adventures', in Hansson, Lindgren Leavenworth, and Ryall, *The Arctic in Literature*, 89–106.
15. Russell Potter focuses on panoramas and engravings in *Arctic Spectacles: The Frozen North in Visual Culture 1818–1875* (Seattle and London: University of Washington Press, 2007) and Robert G. David gives more attention to the

metropolitan Arctic in *The Arctic in the British Imagination 1818–1914* (Manchester: Manchester University Press, 2000). Both books are excellent resources for the study of the visual Arctic in the nineteenth century, although the broad time span covered by each does not allow for in-depth analysis of individual works. Laurie Garrison gives an excellent account of the subjective responses possible to an Arctic panorama in 'Virtual Reality and Subjective Responses: Narrating the Search for the Franklin Expedition through Robert Burford's Panorama', *Early Popular Visual Culture* 10, no. 1 (February 2012): 7–22. Both Ian MacLaren and Constance Martin prioritise lithographs: Ian S. MacLaren, 'The Aesthetic Map of the North, 1845–1859', *Arctic* 28, no. 2 (June 1985): 89–103; Constance Martin, '"Toward no Earthly Pole": The Search for Franklin in Nineteenth Century Art', *Inter-Nord* 19 (1990): 87–100.

16. It is important to acknowledge that this book draws primarily on records created by Europeans, as opposed to Indigenous Arctic peoples, making the narrative necessarily a one-sided one.
17. Richard Harding, *Modern Naval History: Debates and Prospects* (London: Bloomsbury Academic, 2016), 124; 27.
18. Isobel Armstrong, *Victorian Glassworlds: Glass Culture and the Imagination 1830–1880* (Oxford: Oxford University Press, 2008), 336.
19. Margaret Topping, 'Travel Writing and Visual Culture', in *The Routledge Companion to Travel Writing*, ed. Carl Thompson (London: Routledge, 2016), 78–88 at p. 79.
20. Julia Thomas, *Pictorial Victorians: The Inscription of Values in Word and Image* (Athens: Ohio University Press, 2004), 8.
21. Kate Flint, *The Victorians and the Visual Imagination* (Cambridge: Cambridge University Press, 2000), 1.
22. Ibid., 23.
23. A newspaper article in 1855 noted that the practice of filling windows with lithographs was extensive. 'The Art of Lithography', *Morning Chronicle*, 18 January 1855.
24. Richard C. Sha, *The Visual and Verbal Sketch in British Romanticism* (Philadelphia: University of Pennsylvania Press, 1998), 28.
25. Geoff Quilley, '"By Cruel Foes Oppress'd": British Naval Draughtsmen in Tahiti and the South Pacific in the 1840s', *Journal of Historical Geography* 43 (January 2014): 71–84 at pp. 71–2.
26. Richard Sha explains the association of sketches with authenticity. *Visual and Verbal Sketch*, 4.
27. Michael E. Leek, *The Art of Nautical Illustration* (Royston: Quantum Books, 2002), 81.
28. Barry Venning, 'Turner's Whaling Subjects', *The Burlington Magazine* 127, no. 983 (February 1985): 75–83.

29. For example, see MacLaren, 'Aesthetic Map', 89–103; Martin, 'No Earthly Pole', 87–100; Spufford, *Some Time*, 58; David, *British Imagination*, 12.
30. MacLaren, 'Aesthetic Map', 90–1; 101.
31. As a philosophical concept, the sublime is attributed to Longinus, a first- or third-century AD philosopher. Initially intended to refer to a moment of sudden elevation in discourse, it alluded to moments of greatness. Malcolm Heath, 'Longinus and the Ancient Sublime', in *The Sublime: From Antiquity to the Present*, ed. Timothy M. Costello (Cambridge: Cambridge University Press, 2012), 11–23 at p. 23. The ideals of the formal picturesque aesthetic developed from seventeenth-century Italian landscape painting. Malcolm Andrews, *The Idea of the Picturesque and the Vogue for Scenic Tourism*, vol. 1 of *The Picturesque: Literary Sources and Documents*, ed. Graham Clarke (Mountfield: Helm Information, 1994), 3–6.
32. Edmund Burke, *A Philosophical Enquiry into the Origin of Our Ideas of the Sublime and the Beautiful*, ed. Adam Phillips (Oxford: Oxford University Press, 1990), p. xxi.
33. Ibid., 53; 77.
34. Ibid., xxii.
35. Melissa McBay Merrit, 'The Moral Source of the Kantian Sublime', in Costello, *The Sublime*, 37–49 at p. 37.
36. Thomas Weiskel, *The Romantic Sublime: Studies in the Structure and Psychology of Transcendence* (Baltimore: Johns Hopkins University Press, 1976), 38–9.
37. Uvedale Price, *An Essay on the Picturesque, as Compared with the Sublime and the Beautiful; and, on the Use of Studying Pictures, for the Purpose of Improving Real Landscape* (London: Robson, 1796), 61.
38. Ibid., 82.
39. Ibid., 26–7; 70.
40. For example, William Combe and Thomas Rowlandson, *The Tour of Doctor Syntax in Search of the Picturesque* (London: Ackermann, 1812). New editions of the book were published many times during the nineteenth century.
41. William Gilpin published recommendations for sketching British landscape from 1748 to 1792. Andrews, *Picturesque*, 11–12.
42. David Cordingly, *Marine Painting in England 1700–1900* (London: Studio Vista, 1974), 12; 135.
43. David, *British Imagination*, 18–19; Heidi Hansson, 'The Gentleman's North: Lord Dufferin and the Beginnings of Arctic Tourism', *Studies in Travel Writing* 13, no. 1 (February 2009): 61–73 at p. 61.
44. Heidi Hansson, 'Introduction: The Arctic in Literature for Children and Young Adults', in Hansson, Lindgren Leavenworth, and Ryall, *The Arctic in Literature*, 5–26 at p. 5.

45. Such misconceptions have become apparent to the author in response to this research. In fact, over four million people live above the Arctic Circle.
46. Adriana Craciun, *Writing Arctic Disaster: Authorship and Exploration* (Cambridge: Cambridge University Press, 2016), 9.
47. Stefan Jonsson, 'Where Humanism Finds Its Ends: Lessons from Pia Arke and Katarina Pirak Sikku on the Difficulty of Narrating the Arctic', *Studies in Travel Writing* 20, no. 3 (September 2016): 226–36 at p. 229.
48. Spufford, *Some Time*, 188.
49. David, *British Imagination*, 47–8; 50.
50. It is important to note that the map in Figure 0.1 does not show the full extent of Indigenous settlement, only that which was closest to the routes and winter quarters of the maritime search expeditions. There were, and still are, many more Indigenous communities in the Arctic.
51. Palaeo-Inuit cultures had spread across the Bering Strait to Greenland from around 2500 BC; prior to the Franklin searches, contact between Inuit and Europeans had been happening for centuries through exploration, trading, and whaling. Olive Patricia Dickason, *Canada's First Nations: A History of Founding Peoples from the Earliest Times* (Oxford: Oxford University Press, 2002), 56; 67.
52. Richard Vaughan, *The Arctic: A History* (Stroud: Sutton Publishing, 1999), 20–2.
53. William Hulme Hooper, *Ten Months among the Tents of the Tuski, with Incidents of an Arctic Boat Expedition in Search of Sir John Franklin, as far as the Mackenzie River, and Cape Bathurst* (London: John Murray, 1853).
54. Sera-Shriar discusses the ethnologist and surgeon Richard King, who had travelled in the Arctic between 1833 and 1835 and published a 'detailed and sensitive' three-part series on Inuit in 1848. One of King's objectives in doing so was to correct the errors of others who had published, particularly in relation to the Inuit 'Intellectual Character'. 'Arctic Observers: Richard King, Monogenism and the Historicisation of Inuit through Travel Narratives', *Studies in History and Philosophy of Biological and Biomedical Sciences* 51 (2015): 23–31 at pp. 28–9.
55. Mary Louise Pratt, *Imperial Eyes: Travel Writing and Transculturation* (London: Routledge, 1992), 7.
56. Majid Yar, 'Panoptic Power and the Pathologisation of Vision: Critical Reflections on the Foucauldian Thesis', *Surveillance and Society* 1, no. 3 (2003): 254–71 at pp. 260–1.
57. Loomis, 'Arctic Sublime', 109.
58. Invented traditions and place are explored in Karen Till, 'Neotraditional Towns and Urban Villages: The Cultural Production of a Geography or

"Otherness"', *Environment and Planning D: Society and Space* 11, no. 6 (1993): 709–32 at p. 710.

59. These local names survive in the unpublished written records of expedition members and do not appear on official maps.
60. The term *topophilia* 'couples sentiment with place'. Yi-Fu Tuan, *Topophilia: A Study of Environmental Perception, Attitudes, And Values*, Morningside ed. (New York: Columbia University Press, 1990), 113.
61. These are *locale* (a setting for everyday life), *location* (a node linking the place to wider networks), and *sense of place* (identification with a place as distinctive and connected to personal identity). John A. Agnew, *Place and Politics in Modern Italy* (Chicago: University of Chicago Press, 2002), 16.
62. Pratt, *Imperial Eyes*, 201.
63. An imagined geography is the knowledge of a space through images and texts. Joanne P. Sharp, *Geographies of Postcolonialism: Spaces of Power and Representation* (London: Sage, 2009), 12.
64. Rob Shields, *Places on the Margin: Alternative Geographies of Modernity* (London: Routledge, 1991), 163.
65. William Barr, *Arctic Hell-Ship: The Voyage of HMS Enterprise 1850–55* (Edmonton: University of Alberta Press, 2007).
66. Lawrence A. Palinkas and Peter Suedfeld stress 'studying positive aspects is not intended to deny the existence of negative ones, but assessment of either alone results in an incomplete and therefore inaccurate picture of polar service'. Positive effects include enjoyment of the natural environment, the balance of novelty and familiarity, free time, and salutogenic after-effects such as a sense of personal achievement, resoluteness, resiliency, and intimacy with fellow crew members. 'Psychological Effects of Polar Expeditions', *Lancet* 371 (January 2008): 153–63 at pp. 158–9.
67. Model sledges and lyrics of 'Arctic jingling' were among the 'coveted' items that former expedition members gave each other. James Nelson to William T. Mumford, 9 September [no year given], in W. T. Mumford, Private Journal of an Expedition to the Arctic Regions to Ascertain the Fate of Sir John Franklin . . . in the Years 1852. 3. 4 [1 March 1852–24 October 1854], W. T. Mumford Fonds [1852–97], MG24-H80, LAC, Ottawa.
68. Palinkas and Suedfeld, 'Psychological Effects', 158.
69. Francis McClintock, *The Voyage of the 'Fox' in the Arctic Seas: A Narrative of the Discovery of the Fate of Sir John Franklin and His Companions* (London: John Murray, 1859), p. xxii.
70. Vivid accounts of dressmaking and theatricals are found in [William Chimmo], *Euryalus; Tales of the Sea, a Few Leaves from the Diary of a Midshipman* (London: J.D. Potter, 1860).

71. Julie F. Codell, 'The Art of Transculturation', in *Transculturation in British Art, 1770–1930*, ed. Julie F. Codell (London: Routledge, 2016), 1–17 at p. 12.
72. William J. Mills, *Exploring Polar Frontiers: A Historical Encyclopedia*, vol. 1 (Santa Barbara: ABC-CLIO, 2003), 417.
73. Cresswell was in the Baltic when the lithographs were being produced for publication. Advert, *Publishers' Circular*, 1 August 1854; Samuel Gurney Cresswell, *War, Ice and Piracy: The Remarkable Career of a Victorian Sailor: The Journals and Letters of Samuel Gurney Cresswell*, ed. Dominick Harrod (London: Chatham Publishing, 2000), 118.
74. Books that use *Critical Position of H.M.S. Investigator* for their covers include Johann Miertsching, *Frozen Ships: The Arctic Diary of Johann Miertsching 1850–1854*, trans. L. H. Neatby (Toronto: Macmillan, 1967); L. H. Neatby, *The Search for Franklin* (Worthing: Littlehampton Book Services, 1970); Cresswell, *War, Ice and Piracy* (2000); Ken McGoogan, *Fatal Passage: The Untold Story of Scotsman John Rae, the Arctic Adventurer Who Discovered the Fate of Franklin* (London: Bantam, 2002); Edward Adams, *Into the Savage Land: The Alaskan Journal of Edward Adams*, ed. Ernest Sipes (Surrey, BC: Hancock House, 2007); Martin W. Sandler, *Resolute: The Epic Search for the Northwest Passage and John Franklin, and the Discovery of the Queen's Ghost Ship* (New York: Sterling, 2008); Brian Payton, *The Ice Passage: A True Story of Ambition, Disaster, and Endurance in the Arctic Wilderness* (Toronto: Anchor Canada, 2010).
75. Anna Green, *Cultural History* (Basingstoke: Palgrave Macmillan, 2008), 4.
76. Susan Close, 'The Camera and the Contact Zone: Re-Envisioning the Representation of Aboriginal Women in the Canadian North', in Codell, *Transculturation in British Art*, 111–24 at p. 112.
77. Susanne von Falkenhausen, *Beyond the Mirror: Seeing in Art History and Visual Culture Studies*, trans. Nicholas Grindell (Bielefeld: Transcript, 2020), 11.
78. Ludmilla Jordanova, 'Image Matters', *Historical Journal* 51, no. 3 (2008): 777–91 at p. 778.
79. Where this was not possible, high-quality photographs were obtained from the institution.
80. Jordanova, 'Image Matters', 779.
81. Elizabeth Edwards, *Raw Histories: Photographs, Anthropology and Museums* (Oxford: Berg, 2001), 2.
82. Ibid., 3.
83. Ibid., 12.
84. Felix Driver, *Geography Militant: Cultures of Exploration and Empire* (Oxford: Blackwell, 2001), 9.
85. Ibid., 8.

86. Bernard Smith, *Imagining the Pacific: In the Wake of the Cook Voyages* (New Haven: Yale University Press, 1992), 53–4.
87. Leila Koivunen, *Visualizing Africa in Nineteenth-Century British Travel Accounts* (London: Routledge, 2008), 11.
88. Library of Congress, *Thesaurus for Graphic Materials* (Washington, DC: Prints and Photographs Division, 2010), www.loc.gov/pictures/collection/tgm/item/tgm008237/.
89. Bamber Gascoigne, *How to Identify Prints* (London: Thames and Hudson, 1998), 1a.
90. Ibid., 1b.
91. Ibid., 1c.
92. Ian Chilvers (ed.), *The Concise Oxford Dictionary of Art and Artists* (Oxford: Oxford University Press, 2003), 172.
93. Henry Huntington and Gunter Weller, 'Introduction', in *Arctic Climate Impact Assessment*, ed. Carolyn Symon (Cambridge: Cambridge University Press, 2005), 1–20 at p. 2.
94. Charles Emmerson, *The Future History of the Arctic: How Climate, Resources and Geopolitics are Reshaping the North and Why It Matters to the World* (London: Vintage, 2011), 3–4.
95. Huntington and Weller, 'Introduction', 2.
96. Michael Byers, *Who Owns the Arctic? Understanding Sovereignty Disputes in the North* (Vancouver: Douglas and McIntyre, 2009), 109–10.
97. Qikiqtani Inuit Association, *Qikiqtani Truth Commission: Community Histories 1950–1975: Resolute Bay* (Iqaluit: Inhabit Media, 2013), 9.
98. Michael Durey, 'Exploration at the Edge: Reassessing the Fate of Sir John Franklin's Last Arctic Expedition', *The Great Circle* 30, no. 2 (2008): 3–40 at p. 34.
99. P. R. Dawes and R. L. Christie, 'Geomorphic Regions', in *Geology of the Innuitian Orogen and Arctic Platform of Canada and Greenland*, ed. H. P. Trettin (Ottawa: Geological Survey of Canada, 1991), 40–4.
100. J. M. Bumsted, *The Peoples of Canada: A Pre-Confederation History* (Toronto: Oxford University Press, 1992), 203–5.
101. Bumsted, *Peoples of Canada*, 319.
102. Durey, 'Exploration at the Edge', 24.
103. Nanna Katrine Lüders Kaalund, 'From Science in the Arctic to Arctic Science: A Transnational Study of Arctic Travel Narratives, 1818–1883' (PhD thesis, York University, Toronto, 2017), 184.
104. Janice Cavell, *Tracing the Connected Narrative: Arctic Exploration in British Print Culture 1818–1860* (Toronto: University of Toronto Press, 2008), 219. Subsequently, more searches for evidence took place; in 2008, Parks Canada began sonar scanning near King William Island and

located the *Erebus* and the *Terror* in 2014 and 2016. Pertinent sources for the Franklin mystery are: Owen Beattie and John Geiger, *Frozen in Time: The Fate of the Franklin Expedition* (London: Bloomsbury, 1987); Durey, 'Exploration at the Edge' (2008); Janice Cavell, 'Going Native in the North: Reconsidering British Attitudes during the Franklin Search, 1848–59', *Polar Record* 45 (2009): 25–35; Lambert, *Franklin: Tragic Hero* (2009); S. Mays, G. J. R. Maat, and H. H. Boer, 'Scurvy as a Factor in the Loss of the 1845 Franklin Expedition to the Arctic: A Reconsideration', *International Journal of Osteoarchaeology* 25 (2015): 334–44; R. S. Taichman, T. Gross, and M. P. Maceachern, 'A Critical Assessment of the Oral Condition of the Crew of the Franklin Expedition', *Arctic* 70, no. 1 (March 2017): 25–36. Garth Walpole gives an excellent account of the recovery of artefacts from the expedition in *Relics of the Franklin Expedition: Discovering Artifacts from the Doomed Arctic Voyage of 1845*, ed. Russell Potter (Jefferson: McFarland, 2017).

105. Loomis, 'Arctic Sublime', 109.
106. Ross, 'Franklin Search', 65.
107. Erika Behrisch Elce (ed.), *As Affecting the Fate of my Absent Husband: Selected Letters of Lady Franklin Concerning the Search for the Lost Franklin Expedition, 1848–1860* (Montreal and Kingston: McGill-Queen's University Press, 2009), 4.
108. The Admiralty offered double pay to participants, once the ship passed north of the Arctic Circle.
109. William Henry Browne, *Ten Coloured Views Taken during the Arctic Expedition* (London: Ackermann, 1850); Samuel Gurney Cresswell, *A Series of Eight Sketches in Colour . . . of the Voyage of H.M.S. 'Investigator'* (London: Day and Ackermann, 1854); Walter William May, *A Series of Fourteen Sketches Made during the Voyage up Wellington Channel* (London: Day and Son, 1855).
110. 'Panorama of the Arctic Regions', *Observer*, 10 February 1850.

1 'On the Spot'

1. Edward Belcher, *Narrative of the Voyage of H. M. S. Samarang: during the Years 1843–46: Employed Surveying the Islands of the Eastern Archipelago*, 2 vols. (London: Reeve, Benham, and Reeve, 1848).
2. William Henry Browne, *Ten Coloured Views Taken during the Arctic Expedition* (London: Ackermann, 1850); 'Sir James Ross' Arctic Expedition', *Illustrated London News*, 17 November 1849.
3. William Henry Browne, *Rough Sketch Moored to an Iceberg* [1848], N: 1993, SPRI, Cambridge.

4. An interesting parallel can be discerned with women's sketches in the Romantic period. Richard Sha observes that many of them are unsigned: 'the ultimate indication of self-effacement, women's sketches virtually refuse to stand as self-contained works of art'. *The Visual and Verbal Sketch in British Romanticism* (Philadelphia: University of Pennsylvania Press, 1998), 93.
5. Robert G. David, *The Arctic in the British Imagination 1818–1914* (Manchester: Manchester University Press, 2000); Diana Donald, 'The Arctic Fantasies of Edwin Landseer and Briton Riviere: Polar Bears, Wilderness and Notions of the Sublime', *Tate Papers* 13 (April 2010), www.tate.org.uk/research/publications/tate-papers/arctic-fantasies-edwin-landseer-and-briton-riviere-polar-bears; Laurie Garrison, 'Virtual Reality and Subjective Responses: Narrating the Search for the Franklin Expedition through Robert Burford's Panorama', *Early Popular Visual Culture* 10, no. 1 (February 2012): 7–22; Ian S. MacLaren, 'The Aesthetic Map of the North, 1845–1859', *Arctic* 38, no. 2 (June 1985): 89–103; Constance Martin, '"Toward No Earthly Pole": The Search for Franklin in Nineteenth Century Art', *Inter-Nord* 19 (1990): 87–100; Russell Potter, *Arctic Spectacles: The Frozen North in Visual Culture 1818–1875* (Seattle and London: University of Washington Press, 2007).
6. Geoff Quilley, '"By Cruel Foes Oppress'd": British Naval Draughtsmen in Tahiti and the South Pacific in the 1840s', *Journal of Historical Geography* 43 (January 2014): 71–84 at p. 84.
7. Felix Driver, 'Material Memories of Travel: The Albums of a Victorian Naval Surgeon', *Journal of Historical Geography* 69 (July 2020): 32–54 at p. 33.
8. Felix Driver, 'Distance and Disturbance: Travel, Exploration and Knowledge in the Nineteenth Century', *Transactions of the Royal Historical Society* 14 (December 2004): 73–92 at p. 86.
9. Nanna Katrine Lüders Kaalund, 'From Science in the Arctic to Arctic Science: A Transnational Study of Arctic Travel Narratives, 1818–1883' (PhD thesis, York University, Toronto, 2017), 180.
10. In the popular historical novel by Rudy Wiebe, *A Discovery of Strangers* (1994), the character of Robert Hood struggles to aesthetically tame the landscape and frame his sketches as picturesque entities. Susan Birkwood, 'From "Naked Country" to "Sheltering Ice": Rudy Wiebe's Revisionist Treatment of John Franklin's First Arctic Narrative', *Nordlit* 23 (2008): 25–38 at p. 35.
11. We are fortunate to have a library catalogue in existence that details the collection on an Arctic search ship. Henry Briant, 'A Catalogue of the Library Established on board HMS Assistance' [1853], Arctic Pamphlets vol. 3, mg N07/08K, RGS, London.

12. The instructions given to Franklin on 5 May 1845 stated: 'on your arrival in England you are immediately to repair to this office, in order to lay before us a full account of your proceedings in the whole course of your voyage, taking care before you leave the ship to demand from the officers, petty officers, and all other persons on board, the logs and journals they may have kept, together with any drawings or charts they may have made, which are all to be sealed up, and you will issue similar directions to Captain Crozier and his officers. The said logs, journals or other documents to be thereafter disposed of as we may think proper to determine.' British Parliament, 'Papers and Correspondence relative to the Arctic Expedition under Sir John Franklin', in *Arctic Blue Books Online: British Parliamentary Papers on Exploration in the Canadian North*, 1818–1878, indexed by Andrew Taylor (Winnipeg: University of Manitoba, 2003), 1848a, 7, www.umanitoba.ca/libraries/units/archives/collections/subject/arcticstudies/arcticbb/.
13. Trevor Harvey Levere, *Science and the Canadian Arctic: A Century of Exploration 1818–1918* (Cambridge: Cambridge University Press, 1993), 173.
14. Claudio Greppi, '"On the Spot": Traveling Artists and the Iconographic Inventory of the World, 1769–1859', in *Tropical Visions in an Age of Empire*, ed. Felix Driver and Luciana Martins (Chicago: University of Chicago Press, 2005), 23–42 at p. 28.
15. Driver, 'Distance and Disturbance', 88.
16. The first book may be R. P. Noble, *A Guide to Water Colour Painting, with an Illustration in Colour* (London: George Rowney, 1850). The second is George Field, *Rudiments of the Painters' Art; or a Grammar of Colouring, Applicable to Operative Painting, Decorative Architecture, and the Arts* (London: John Weale, 1850).
17. Briant, 'Catalogue', 7. It is likely that many of the same books were in the library when the *Assistance* formed part of the Austin expedition from 1850 to 1851. The catalogue was printed on board and done in a manner that makes it difficult to count the number of books and periodicals with exactitude.
18. These books were probably Robert Hunt, *Photography: A Treatise on the Chemical Changes Produced by Solar Radiation, and the Production of Pictures from Nature, by the Daguerreotype, Calotype, and Other Photographic Processes* (London: Griffin, 1851); William Thornthwaite, *A Guide to Photography, Containing Simple and Practical Details* (London: Horn, Thornthwaite, and Wood, 1845).
19. From my study of Briant's catalogue, the estimated percentages by subject area of the contents of the *Assistance*'s library are as follows: Geography 26%; History 20%; Literature, Languages, and the Arts 20%; Science 13%; Religion 10%; Miscellaneous 9%; Periodicals 2%. These figures are based on my own

calculations. Fictional narratives such as *Gulliver's Travels* also appear in the catalogue.

20. Alexander von Humboldt, 'Landscape Painting', in *Cosmos: Sketch of a Physical Description of the Universe*, vol. 2, trans. under the superintendence of Edward Sabine (London: Longman, Brown, Green, and Longmans and John Murray, 1849), 74–91.
21. Ibid., 87.
22. Ibid., 74.
23. Nigel Leask, *Curiosity and the Aesthetics of Travel Writing, 1770–1840* (Oxford: Oxford University Press, 2002), 248–9.
24. Levere, *Science and the Canadian Arctic*, 142; Andrew Lambert, *Franklin: Tragic Hero of Polar Navigation* (London: Faber & Faber, 2009), 192–9.
25. Bernard Lightman, 'The Visual Theology of Victorian Popularizers of Science: From Reverent Eye to Chemical Retina', *Isis* 91, no. 4 (December 2000): 651–80 at pp. 653–5.
26. Ibid., 658.
27. Moshe Barasch, *Modern Theories of Art: From Winckelmann to Baudelaire* (London: New York University Press, 1990), 246.
28. Luciana Martins, 'The Art of Tropical Travel, 1768–1830', in *Georgian Geographies: Essays on Space, Place and Landscape in the Eighteenth Century*, ed. Miles Ogborn and Charles W. J. Withers (Manchester: Manchester University Press, 2004), 72–91 at pp. 73; 75.
29. Lindsay Stainton, *Nature into Art: English Landscape Watercolours* (London: British Museum Press, 1991), 13.
30. Marcia Pointon, 'Geology and Landscape Painting in Nineteenth-Century England', in *Images of the Earth*, 2nd ed., ed. Ludmilla Jordanova and Roy Porter (London: British Society for the History of Science, 1997), 93–123 at pp. 97–8.
31. Martins, 'Tropical Travel', 81.
32. Erika Behrisch Elce, '"Adapted for Travellers in General": En Route with the British Admiralty's Manual of Scientific Enquiry', *Journeys* 15, no. 1 (2014): 1–22 at p. 6.
33. Briant, 'Catalogue', 1.
34. John F. W. Herschel, ed., *A Manual of Scientific Enquiry: Prepared for the Use of Her Majesty's Navy and Adapted for Travellers in General. Published by Authority of the Lords Commissioners of the Admiralty* (London: John Murray, 1849), p. i.
35. Levere, *Science and the Canadian Arctic*, 144.
36. Herschel, *Admiralty Manual*, p. iii.
37. Charles Darwin, 'Geology', in Herschel, *Admiralty Manual*, 180.
38. Louis Agassiz, *Études sur les Glaciers* (Neuchatel: Jent and Gassmann, 1840).

39. William J. Hamilton, 'Geography', in Herschel, *Admiralty Manual*, 134.
40. Samuel Gurney Cresswell, *War, Ice and Piracy: The Remarkable Career of a Victorian Sailor: The Journals and Letters of Samuel Gurney Cresswell*, ed. Dominick Harrod (London: Chatham Publishing, 2000), 56.
41. British Parliament, *Arctic Blue Books*, 1850a, 66.
42. Neville W. Poulsom, *British Polar Exploration and Research: A Historical and Medallic Record with Biographies 1818–1999* (London: Savannah Publications, 2000), 59.
43. James J. Rutter, *A View of the Three Glaciers at the Eastern Extremity of Wolstenholme Sound* [17 June 1850], ADM 344/2020, Admiralty Records, TNA, London. A version of the painting exists in the British Library as part of the Barrow bequest: Add MS 35304, Barrow Bequest, vol. 5, BL. The inscription 'sketched and drawn from nature' does not appear on this version, suggesting that it is a copy of the painting in TNA. The version in the British Library is unsigned.
44. British Parliament, *Arctic Blue Books*, 1851a, 62.
45. Ronald Heus, Hein A. M. Daanen, and George Havenith, 'Physiological Criteria for Functioning of Hands in the Cold', *Applied Ergonomics* 26, no. 1 (1995): 5–13 at pp. 6; 11.
46. While man-hauling is associated with British polar exploration, dogs were also used to pull sledges on the Franklin search expeditions.
47. British Parliament, *Arctic Blue Books*, 1852b, 14.
48. British Parliament, *Arctic Blue Books*, 1852b, 64. The special instruction given to Browne was as a result of his work on the Ross expedition (1848–9), after which his visual records were used as the basis of the panorama *Summer and Winter Views of the Polar Regions* (examined in Chapter 4) and a folio of lithographs *Ten Coloured Views* (examined in Chapter 5).
49. British Parliament, *Arctic Blue Books*, 1852b, 30.
50. Martin J. S. Rudwick, 'The Emergence of a Visual Language for Geological Science 1760–1840', *History of Science* 14 (1976): 149–95 at p. 153.
51. William Henry Browne, *Cape Walker Encampment* [1851], Y: 54/19/9, SPRI, Cambridge.
52. British Parliament, *Arctic Blue Books*, 1852b, 75. Officers were attuned to observing potential sources of coal in the Arctic, the mining of which would avoid having to transport coal from England for the steamers.
53. Kate Flint, *The Victorians and the Visual Imagination* (Cambridge: Cambridge University Press, 2000), 137–8.
54. Barry Lopez, *Arctic Dreams: Imagination and Desire in a Northern Landscape*, 2nd ed. (London: Harvill Press, 1999), 226.
55. John Franklin Aylen, *Perilous Position of HMS North Star, Sept 21st 1849, NE end of Baffin's Bay* [1849], ADM 344/2030, Admiralty Records, TNA, London.

56. British Parliament, *Arctic Blue Books*, 1851a, 57.
57. Add MS 35304, Barrow Bequest, vol. 5, BL.
58. British Parliament, *Arctic Blue Books*, 1850a, 65.
59. Erika Behrisch, '"Far as the Eye Can Reach": Explorers' Poetry in the Arctic 1832–1852', in 'Science and Victorian Poetry', ed. Gowan Dawson and Sally Shuttleworth, special issue, *Victorian Poetry* 41, no. 1 (Spring 2003), 73–92 at p. 73.
60. Behrisch, 'Explorers' Poetry', 85.
61. Behrisch, 'Explorers' Poetry', 77.
62. Richard C. Sha, *The Visual and Verbal Sketch in British Romanticism* (Philadelphia: University of Pennsylvania Press, 1998), 68.
63. Majid Yar, 'Panoptic Power and the Pathologisation of Vision: Critical Reflections on the Foucauldian Thesis', *Surveillance and Society* 1, no. 3 (2003): 254–71 at pp. 260–1.
64. Walter William May, on Belcher's expedition from 1852 to 1854, wrote that he 'attempted to make a sketch' of a particularly beautiful scene. May, Journal, Entry for 30 November 1852. Austin expected his officers to 'make' sketches of headlands. British Parliament, *Arctic Blue Books*, 1852b, 14.
65. Helen M. Rozwadowski, 'Small World: Forging a Scientific Maritime Culture for Oceanography', *Isis* 87 (2005): 409–29 at pp. 424–5.
66. William T. Domville, JOD/67, Private Journal, NMM, Greenwich.
67. Library of Congress, *Thesaurus*, www.loc.gov/pictures/collection/tgm/item/tgm011033/.
68. William T. Domville, *Esquimaux Dance aboard Resolute* [1852], JOD/67, Private Journal, NMM, Greenwich.
69. The dance and the Greenlanders' knowledge of such dances seem to have made an impression on expedition members, and descriptions of it appear in many formats including: Walter W. May, Journal of HMS Assistance [20 April 1852–19 August 1853], Walter May Fonds MG24-H31: Entry for 8 June 1852, LAC, Ottawa; Henry Kellett to John Barrow, 9 June 1852, Add 35307 Sir Edward Belcher's Expedition, Manuscripts, BL; James Blair Grove to John Barrow, 17 June 1852, Add 35307 Sir Edward Belcher's Expedition, Manuscripts, BL; Edward Belcher, *The Last of the Arctic Voyages* (London: Lovell Reeve, 1855), 1: 38; George Frederick McDougall, *The Eventful Voyage of H. M. Discovery Ship 'Resolute' to the Arctic Regions* (London: Longman, Brown, Green, Longmans, & Roberts, 1857), 36; Emile de Bray, *A Frenchman in Search of Franklin: De Bray's Arctic Journal 1852–54*, trans. and ed. William Barr (Toronto: University of Toronto Press, 1992), 23.
70. I am grateful to Sue Jones at the University of Oxford for identifying the dance depicted.

71. Belcher, *Arctic Voyages*, 38; McDougall, *Eventful Voyage*, 36; May, Journal, Entry for 8 June 1852; James Blair Grove to John Barrow, 17 June 1852; Henry Kellett to John Barrow, 9 June 1852.
72. McDougall, *Eventful Voyage*, 27.
73. Ibid., 35. The women were half-Danish and half-Inuit. McDougall, *Eventful Voyage*, 201.
74. Bray, *Frenchman in Search*, 23.
75. McDougall, *Eventful Voyage*, 36.
76. 'I enclose a hurried sketch I took of one of them and got her to write her name underneath it is not a very ladylike hand but still not bad.' James Blair Grove to John Barrow, 17 June 1852, Add MS 35307 Sir Edward Belcher's Expedition, Manuscripts, BL.
77. *Illustrated London News*, 11 December 1852.
78. Robert McCormick to John Barrow, 9 June 1852, Add 35307 Sir Edward Belcher's Expedition, Manuscripts, BL.
79. Johann Miertsching, *Frozen Ships: The Arctic Diary of Johann Miertsching 1850–1854*, trans. L. H. Neatby (Toronto: Macmillan, 1967), 99. It should be noted that Miertsching's 'daily journal was left behind on the ship', and he wrote this version with the aid of McClure's diary, his own notes, and his memory. Miertsching, *Frozen Ships*, 3.
80. Although such unframed pictures could conceivably be rolled up, such an item would still be difficult to conceal and transport in bad weather; they would inevitably have become damaged while hauling a sledge for fourteen days to the *Resolute* at Melville Island. Robert McClure, *The Discovery of the North-West Passage*, ed. Sherard Osborn (London: Longman, Brown, Green, Longmans, & Roberts, 1856), 288.
81. The 1849 watercolour from the Ross expedition is threaded with string, forming a loop at the top, and was found in his journal from Russian America during the winter of 1850 to 1851, which he spent with two other members of the Collinson expedition. The picture shows the two ships *Enterprise* and *Investigator* in the ice of Barrow Strait. Edward Adams, MS 1115, Journal [12 October 1850 to 3 July 1851], SPRI, Cambridge.
82. Robert Dawes Aldrich, *Sailing on the Midnight Deep* [1850–1], PAH0072, NMM, Greenwich; Aldrich, *Morn Alas Will Not Restore Us* [1850–1], PAH0071, NMM, Greenwich. The paintings are catalogued as such, although, being affixed to card (a common occurrence with such material), it is not possible to see if they were signed and dated on the reverse.
83. *A Select Collection of Songs* (Newcastle Upon Tyne, 1806), 6–7; *The Universal Songster; or, Museum of Mirth* (London, 1826), 2:103.

84. Aldrich sang a song of his own composition at the close of the Royal Arctic Theatre on board the *Assistance* on 4 March 1851 to the air of 'Hearts of Oak'. Sherard Osborn and George F. McDougall, *Illustrated Arctic News* (London: Ackermann, 1852), 42.
85. Thomas Haynes Bayly, *The Poetical Works of Thomas Haynes Bayly*, ed. Helena Beecher Haynes Bayly, 2 vols. (London: Richard Bentley, 1844), 1:182–3.
86. William Gerard Don, 'Reminiscences of the Baltic Fleet of 1855' (Brechin, 1894), 124–5.
87. Also known as a mock moon, a paraselene is 'an optical phenomenon resulting from the refraction and reflection of moonlight within ice crystals in cirrus or cirrostratus cloud'. Ian Ridpath, *A Dictionary of Astronomy* (Oxford: Oxford University Press, 2012), 350.
88. Robert Dawes Aldrich, *Paraselene* [1850–1], PAG8004, NMM, Greenwich.
89. Officers and Seamen of the Expedition, *Arctic Miscellanies: A Souvenir of the Late Polar Search*, ed. James Donnet (London: Colburn and Company, 1852), 41.
90. William T. Mumford, *The First View of the Coast Greenland 21.5.52* [1852], ICON5577, LAC, Ottawa.
91. W. T. Mumford, Private Journal of an Expedition to the Arctic Regions to Ascertain the Fate of Sir John Franklin ... in the Years 1852. 3. 4 [1 March 1852–24 October 1854], W. T. Mumford Fonds MG24-H80: Entry for 21 May 1852, LAC, Ottawa. Walter W. May, Journal of HMS Assistance [20 April 1852–19 August 1853], Walter May Fonds MG24-H31: Entry for 11 October 1852, LAC, Ottawa.
92. Richard Harding, *Modern Naval History: Debates and Prospects* (London: Bloomsbury Academic, 2016), 119.
93. William Blakey, *An Incident in the Search for Sir John Franklin* [1854], PAG8039, NMM, Greenwich.
94. James Cook, *A Voyage towards the South Pole and round the World* (London: W. Strahan and T. Cadell, 1777), 36.
95. Edward Adams, Journal [12 October 1850–3 July 1851], MS 1115, SPRI, Cambridge.
96. Edward Adams, *Koutoküdluk – My First Love, 1851* [1851], Y: 83/11/55, SPRI, Cambridge.
97. Kristina Huneault, 'Miniature Objects of Cultural Covenant: Portraits and First Nations Sitters in British North America', *Canadian Art Review* 30, no. 1/2 (2005): 87–100 at pp. 87–8.
98. Adams, Journal, Entry for 11 January 1851.
99. H. Stevenson, 'The Late Mr. Edward Adams', *Ibis, a Quarterly Journal of Ornithology* 2, no. 8 (October 1878): 420–42 at p. 422.

100. Heidi Hansson, 'The Gentleman's North: Lord Dufferin and the Beginnings of Arctic Tourism', *Studies in Travel Writing* 13, no. 1 (February 2009): 61–73 at p. 61.
101. Hansson, 'Gentleman's North', 61.
102. Chauncey C. Loomis, 'The Arctic Sublime', in *Nature and the Victorian Imagination*, ed. U. C. Knoepflmacher and G. B. Tennyson (Berkeley and Los Angeles: University of California Press, 1977), 95–112 at pp. 97–8.
103. Ibid., 101.
104. Loomis, 'Arctic Sublime', 103.
105. Ann Radcliffe, 'The Snow-Fiend', in *Gaston de Blondeville and St. Alban's Abbey*, 4 vols. (London: Henry Colburn, 1826), 4:192–5. Lines from the poem were also quoted in the *London Journal* as a 'perfectly true' description. 'The Arctic Expedition', 24 November 1849.
106. Greppi, '"On the Spot"', 23.
107. The clerk of the *North Star* (1849–50), James Rutter, did write 'on the spot' on his work, possibly because, as a clerk and not a midshipman, he was less familiar with the conventions of naval drawing.
108. Cresswell, *War, Ice and Piracy*, 56.
109. Ibid., 57.
110. Edward Adams, Journal [8 May 1848–29 October 1849], MS 1461:1, Entry for 30 July 1848, SPRI, Cambridge.
111. Walter May Fonds [1852–4], MG24-H31, LAC, Ottawa.
112. Robert McCormick, Journal [28 April–21 June 1852], MS 3374; Journal [19 August–11 September 1852], MS 3375; Journal [August–September 1852], MS 3376; Journal [August–September 1852], MS 3377, Sketchbook, [1852–3], MS 3382, Wellcome Institute, London.
113. W. T. Mumford, Private Journal of an Expedition to the Arctic Regions to Ascertain the Fate of Sir John Franklin … in the Years 1852. 3. 4 [1 March 1852–24 October 1854], W. T. Mumford Fonds MG24-H80, [1852–97], LAC, Ottawa.
114. HMS *Assistance* and HMS *Intrepid* visited the bay; according to Sherard Osborn: 'The fact of the "North Star" having wintered last year in Wolstenholme Sound, or "Petowack", was elicited … The "Assistance" and "Intrepid", therefore remained to visit that neighbourhood … So ended the memorable 14th of August.' Sherard Osborn, *Stray Leaves from an Arctic Journal* (London: Longman, Brown, Green, and Longmans, 1852), 86–7.
115. [Unknown Artist], *North Star Mount in Wolstenholme Sound, SE. 1.8 miles: land of a brownish purplish tinge, August 15th 1850* [1850], T3262 / nla.obj-133007737, NLA, Canberra. It was not possible to view this painting in person, but the National Library of Australia helpfully emailed a digital photograph.

116. The *North Star* had broken out of her winter quarters on 1 August 1850, and Saunders mentions that 'during the latter part of July the ice decayed in an astonishingly rapid manner; the floe, which some short time ago was 4 feet in thickness, was now broken into small pieces, and was fast disappearing, so much that the water alongside the ship on the surface was perfectly fresh'. British Parliament, *Arctic Blue Books*, 1851a, 60.
117. Berthold Seemann, 'Flora of Western Eskimaux-Land', in *The Botany of the Voyage of H.M.S. Herald . . . during the Years* 1845–51 (London: Reeve, 1852), 11–56.
118. Cresswell, *War, Ice and Piracy*, 57. In the nineteenth century, watercolours were sold as hard dry cakes that needed to be rubbed in a small amount of water to create paint.
119. Samuel Gurney Cresswell, *Brown's Island Aug 1848* [1848], WMH 3–1-D4-1, NRO, Norwich.
120. Adams, Journal, Entry for 8 August 1848.
121. Adams, Journal, Entry for 8 July 1848.
122. The set of photographs by Inglefield compiled in an album and presented to Queen Victoria and Prince Albert in 1854 is available to view in high resolution on the Royal Collection Trust website: www.rct.uk/collection/2510457/h-m-ships-phoenix-talbot-in-search-of-sir-john-franklin-1854.
123. William T. Domville, P35CAL, P36CAL, P37CAL, P38CAL, NMM, Greenwich. The calotype (also known as the talbotype) was an early photographic process, patented in 1839 by William Henry Fox Talbot, which competed with the daguerreotype. Asa Briggs, *Victorian Things* (Stroud: Sutton Publishing, 2003), 102. May wrote: 'I have been amusing myself by endeavouring to prepare paper for the Talbotype. But I have not succeeded as well as I should have expected.' May, Journal, Entry for 15 July 1852.
124. Efram Sera-Shriar, 'Anthropometric Portraiture and Victorian Anthropology: Situating Francis Galton's Photographic Work in the Late 1870s', *History of Science* 53, no. 2 (2015): 155–79 at p. 159.
125. Briggs, *Victorian Things*, 107.
126. Edward Augustus Inglefield, *The Church and Parsonage, Holsteinborg* [1854], G4271, NMM, Greenwich.
127. Edward Augustus Inglefield, *Unidentified Elderly Inuit Woman* [1854], G4269, NMM, Greenwich.
128. Sera-Shriar, 'Anthropometric Portraiture', 159.
129. May, Journal, Entry for 30 November 1852. Here, May has left a blank space for the word beginning with 'P'. Perhaps he wanted to check the term or its spelling before entering it in his journal. When May mentions bringing out

his drawings, he is referring to the illustrations for the 'Queen's Illuminated Magazine'. In fact, the picture appears on the title page of the second issue (December 1852). The oval-shaped picture is painted in blue and white and affixed to the magazine.

130. Osborn, *Stray Leaves*, 168–9.

131. Edward Adams, *Glacier Bay, copied from Lieut Brown by candle light* [1848–9], Y: 83/11/12, SPRI, Cambridge.

2 'Breathing Time'

1. Walter W. May, Journal of HMS Assistance [20 April 1852–19 August 1853], Walter May Fonds MG24-H31: Entry for 2 December 1852, LAC, Ottawa. See also entries for 21, 28 October; 2, 8, 30 November; 14, 20, 21 December 1852. Dissolving views involved the use of two of more lanterns so that an image could be superimposed on another similar image, which would then fade. This gave the impression of the uninterrupted transformation of a scene. Kevin Rockett and Emer Rockett, *Magic Lantern, Panorama and Moving Picture Shows in Ireland, 1786–1909* (Dublin: Four Courts Press, 2011), 42.
2. Sherard Osborn, 'Queen's Illuminated Magazine and North Cornwall Gazette. Published in Winter Quarters, Arctic Regions' [28 October 1852–12 February 1853]: Add MS 35305, Barrow Bequest, vol. 6, Western Manuscripts, BL. Although the full title indicated it was 'published', this referred to its availability for both ships' companies in winter quarters. The ink illustrations were drawn directly onto the manuscript while the watercolour paintings were affixed to it.
3. It is not known if the original manuscript exists, but a facsimile of the 'Illustrated Arctic News' was engraved and published when the expedition returned. The facsimile is discussed throughout the chapter: Osborn and McDougall, *Illustrated Arctic News*, 1852. Each page of the *Illustrated Arctic News* can be viewed online on the Royal Museums Greenwich website: https://collections.rmg.co.uk/collections/objects/137953.html.
4. While the term 'illustrated' implies text accompanied by visual material in either black and white or colour, the term 'illuminated' specifically indicates colour as a feature.
5. 'A Visit to the Arctic Ships', *Chambers's Edinburgh Journal*, 25 May 1850.
6. William Barr, *Arctic Hell-Ship: The Voyage of HMS Enterprise 1850–55* (Edmonton: The University of Alberta Press, 2007); Andrew Lambert, *The Gates of Hell: Sir John Franklin's Tragic Quest for the North West Passage* (New Haven: Yale University Press, 2009); Ken McGoogan, *Fatal Passage: The Untold Story of Scotsman John Rae, the Arctic Adventurer who Discovered the*

Fate of Franklin (London: Bantam, 2002). It should be noted that Lambert's book was titled *Franklin: Tragic Hero of Polar Navigation* in the UK (London: Faber & Faber, 2009).

7. Officers and Seamen of the Expedition, *Arctic Miscellanies: A Souvenir of the Late Polar Search*, ed. James Donnet (London: Colburn and Co., 1852); Sherard Osborn and George F. McDougall, *Illustrated Arctic News* (London: Ackermann, 1852).
8. Henry Briant, 'A Catalogue of the Library Established on board HMS Assistance' [1853], Arctic Pamphlets vol. 3, mg N07/08K, RGS, London. The entries in the catalogue suggest that bound volumes of periodicals were on board, although this is not certain.
9. Elaine Hoag, 'Shipboard Printing on the Franklin Search Expeditions: A Bibliographical Study of the *Plover* Press', *Papers of the Bibliographical Society of Canada* 38, no. 1 (2000): 7–71 at p. 28.
10. Elaine Hoag, 'Caxtons of the North: Mid-Nineteenth-Century Arctic Shipboard Printing', *Book History*, 4 (2001): 81–114 at pp. 99; 85–6.
11. David H. Stam and Deirdre C. Stam, 'Bending Time: The Function of Periodicals in Nineteenth-Century Polar Naval Expeditions', *Victorian Periodicals Review* 41, no. 4 (Winter 2008): 301–22 at pp. 302; 304.
12. Erika Behrisch Elce, '"One of the Bright Objects that Solace Us in these Regions": Labour, Leisure, and the Arctic Shipboard Periodical, 1820–1852', *Victorian Periodicals Review* 46, no. 3 (Fall 2013): 343–67 at p. 344.
13. Ibid., 347.
14. Ibid., 352; 358.
15. Hester Blum, 'The News at the Ends of the Earth: Polar Periodicals', in *Unsettled States: Nineteenth-Century American Literary Studies*, ed. Dana Luciano and Ivy G. Wilson (New York: New York University Press, 2014), 158–88 at p. 165.
16. Ibid., 165.
17. Ibid., 161.
18. Hester Blum, *The News at the Ends of the Earth: The Print Culture of Polar Exploration* (Durham, NC: Duke University Press, 2019), 28; 138.
19. Colburn & Co. Publishers, handbill advertising *Arctic Miscellanies. A Souvenir of the Late Polar Search* [1852], Erasmus Ommanney Collection EO/3/3, RGS, London.
20. Edward Sabine (ed.), 'The New Georgia Gazette or Winter Chronicle', [1819–1820] MS 438/12. SPRI, Cambridge. The manuscript was subsequently published as part of Parry's *Journal of a Voyage for the Discovery of a North-West Passage* (London: John Murray, 1821).
21. Blum, 'Polar Periodicals', 171.

22. John Matthews, The Journal of John Matthews from September 1850 to April 1855, written during a voyage in search of the Franklin Expedition, [1850–5] mg No7/11D: Entry for November 1852. Monograph Collection, RGS, London.
23. Blum, 'Polar Periodicals', 162.
24. *Arctic Miscellanies*, p. ix.
25. Heather Davis-Fisch, *Loss and Cultural Remains in Performance: The Ghosts of the Franklin Expedition* (New York: Palgrave Macmillan, 2012); Heidi Hansson, 'Staging the Arctic 1819–1909 and 2014', *Nordlit* 35 (2015): 47–62; Moira Dunbar, 'The Royal Arctic Theatre, with Drawings from The Illustrated Arctic News published on board H.M.S. Resolute in the Canadian Arctic, 31 October 1850 to 14 March 1851', *Canadian Art* 15, no. 2 (1958): 110–113; Patrick B. O'Neill, 'Theatre in the North: Staging Practices of the British Navy in the Canadian Arctic', *Dalhousie Review* 74, no. 3 (1994): 356–84.
26. Phyllis J. Johnson and Peter Suedfeld, 'Coping with Stress through the Microcosms of Home and Family Among Arctic Whalers and Explorers', *History of the Family* 1, no. 1 (1996): 41–62 at pp. 42–3.
27. Lawrence A. Palinkas and Peter Suedfeld, 'Psychological Effects of Polar Expeditions', *Lancet* 371 (January 2008): 153–63 at p. 158.
28. Osborn, 'Queen's Illuminated Magazine', 1: 1.
29. Carolyn Strange, 'Reconsidering the "Tragic" Scott Expedition: Cheerful Masculine Home-making in Antarctica, 1910–1913', *Journal of Social History* 48, no. 1 (Fall 2012): 66–88 at p. 73.
30. Marcus Rediker, *Between the Devil and the Deep Blue Sea*, Canto edition (Cambridge: Cambridge University Press, 1993), 159.
31. Strange, 'Scott Expedition', 74.
32. William Edward Parry established that the practice of amusements would negate boredom and listlessness during the long Arctic winter. Blum, 'Polar Periodicals', 171.
33. Henri Lefebvre, *The Production of Space*, trans. Donald Nicholson-Smith (Oxford: Blackwell, 1991), 86.
34. Rob Shields, *Lefebvre, Love and Struggle: Spatial Dialectics* (London: Routledge, 1999), 158.
35. The 'Polar Almanac for 1854' printed by Henry Hester on board the *Enterprise* is not included here. The forty-page 'Almanac' differs from the periodicals in its intention and recorded more factual information about the voyage. Neither does the figure include periodicals on private expeditions.
36. Hoag, 'Caxtons of the North', 102. The 'Weekly Guy' was a four-page periodical printed on the *Plover* from November 1852 to February 1853

when the ship was under the command of Rochfort Maguire. John Simpson, ed. 'Weekly Guy' [1852–3], Arctic Pamphlets vol. 3, item 7 rgs302369/N07/08K, RGS, London.

37. Hoag, 'Caxtons of the North', 102; May, Journal, Entry for 11 October 1852. Printing presses were first brought on board for printing messages to be distributed in the Arctic, but were also used to print playbills, songs, and announcements.
38. May, Journal, Entry for 11 October 1852.
39. John Bertie Cator, Article from the Arctic Charivari [6 August 1851], Erasmus Ommanney Collection EO/3/3, RGS, London.
40. John Simpson, 'Flight of the Plover' [1848], John Simpson Papers RL.01195, Box 1, RL, Duke University, North Carolina.
41. Not enough information exists on two of the periodicals that do not survive ('Gleaner' and 'Arctic Charivari'). I am grateful to Mary Caton Lingold at the Rubenstein Rare Book and Manuscript Library for kindly scanning and emailing 'Flight of the Plover' documents free of charge.
42. Albert Hastings Markham, *The Life of Sir Clements R. Markham, K.C.B., F.R.S.* (London: John Murray, 1917), 120. The original contains more illustrations than the published version.
43. Osborn and McDougall, *Illustrated Arctic News*, 1852; *Arctic Miscellanies*, 1852. The original 'Aurora Borealis' exists in the National Maritime Museum, Greenwich. I was, however, unable to view it in the course of this research.
44. Arthur L. Bowley, *Wages in the United Kingdom in the Nineteenth Century* (Cambridge: Cambridge University Press, 1900), 78.
45. Clements R. Markham, *The Arctic Navy List 1773–1873* (1875; repr. Dallington: Naval and Military Press, 1992), 39.
46. Sherard Osborn to William Penny, 27 September 1856, MS 116/57/1–12, SPRI, Cambridge.
47. Samuel Gurney Cresswell, *War, Ice and Piracy: The Remarkable Career of a Victorian Sailor: The Journals and Letters of Samuel Gurney Cresswell*, ed. Dominick Harrod (London: Chatham Publishing, 2000), 143.
48. Markham, *Arctic Navy List*, 32–3. McDougall also produced some illustrations for 'Aurora Borealis'.
49. John Bertie Cator, Diary HMS Intrepid [1850–51], 2: Entry for 28 May 1851, SPRI, Cambridge.
50. Markham, *Arctic Navy List*, 35.
51. May, Journal, Entry for 6 January 1854.
52. Andrew Lambert, *Tragic Hero*, 230.
53. Markham, *Arctic Navy List*, 35; Christopher Wood, *Dictionary of British Art*, vol. 4, bk. 1, *Victorian Painters* (Woodbridge: Antique Collectors' Club, 1998), 348.

54. Walter William May, *A Series of Fourteen Sketches Made during the Voyage up Wellington Channel* (London: Day and Son, 1855).
55. The 'Queen's Illuminated Magazine' is 34 × 28 cm and the *Illustrated Arctic News* facsimile is 47.5 × 29.3 cm.
56. I have not yet been able to trace the whereabouts of the actual manuscript of the 'Illustrated Arctic News'.
57. This was highly likely to have been due to a breakdown in the social cohesion of the group under Belcher's command.
58. Alvin Sullivan, ed., *British Literary Magazines: The Victorian and Edwardian Age, 1837–1913* (London: Greenwood Press, 1984), p. vii.
59. Ibid., p. viii.
60. The author here implies that the term 'magazine' is associated with a lack of 'news'. Printed advertisement for the 'Queen's Illuminated Magazine and North Cornwall Gazette. Published in Winter Quarters, Arctic Regions' [1852], Add MS 35305, Barrow Bequest, vol. 6, Western Manuscripts, BL.
61. May, *Fourteen Sketches*, 1855. Anonymity was common in journalism prior to the 1860s (Sullivan, *British Literary Magazines*, p. xii), and the entire magazine, both the text and the illustrations, is made up of anonymous contributions, although it is likely that the identity of some of the authors was known.
62. May, Journal, Entry for 11 October 1852.
63. May, Journal, Entry for 8 November 1852.
64. Sullivan, *British Literary Magazines*, p. xvi; Patricia Anderson, 'Illustration', in *Victorian Periodicals and Victorian Society*, ed. J. Don Vann and Rosemary T. VanArsdel (Aldershot: Scolar, 1994), 127–42 at pp. 127–30.
65. Osborn, 'Queen's Illuminated Magazine', 4 [February 1853]: 1.
66. Osborn, 'Queen's Illuminated Magazine', 4 [February 1853]: 1.
67. Osborn, 'Queen's Illuminated Magazine', 4 [February 1853]: 2.
68. Osborn, 'Queen's Illuminated Magazine', 4 [February 1853]: 2.
69. *Arctic Miscellanies*, 202.
70. [William Chimmo], *Euryalus; Tales of the Sea, a Few Leaves from the Diary of a Midshipman* (London: J.D. Potter, 1860), 300–1.
71. Vern L. Bullough and Bonnie L. Bullough, *Cross Dressing, Sex, and Gender* (Philadelphia: University of Pennsylvania Press, 1993), 232–5.
72. Davis-Fisch, *Loss and Cultural Remains*, 33.
73. David A. Boxwell, 'The Follies of War: Cross-Dressing and Popular Theatre on the British Front Lines, 1914–18', *Modernism/Modernity* 9, no. 1 (January 2002): 1–20 at pp. 13–14. Boxwell discusses the divergent traditions of 'mimicry and mimesis' that became apparent in the history of female impersonation.
74. Eve Kosofsky Sedgwick, *Epistemology of the Closet*, 2nd ed. (Berkeley and Los Angeles: University of California Press, 2008), 2.

75. Boxwell, 'Follies of War', 9–10.
76. I deal with these aspects in Chapters 4 and 5.
77. Osborn, 'Queen's Illuminated Magazine', 4 [February 1853]: 3.
78. Ibid., 6.
79. Osborn, 'Queen's Illuminated Magazine', 2 [December 1852]: 8.
80. Clements Markham, Private Journal from HMS Assistance [1850–51], The Sir Clements Robert Markham Collection CRM/3: 33, RGS, London.
81. Clements Markham, *Franklin's Footsteps: A Sketch of Greenland, along the Shores of Which His Expedition Passed, and of the Parry Isles, where the Last Traces of It Were Found* (London: Chapman and Hall, 1853), 79.
82. Richard Altick, *Punch: The Lively Youth of a British Institution* (Columbus: Ohio State University Press, 1997), 6. *Punch* measured 28 × 22 cm, making it smaller than both the 'Queen's Illuminated Magazine' and the *Illustrated Arctic News*.
83. Altick, *Punch*, p. xvii.
84. *Cruikshank's Table-Book* measured 23.8 × 17.1 cm. George Cruikshank, *George Cruikshank's Table-Book*, ed. Gilbert Abbott and A. Beckett (London: Punch, 1845); Briant, 'Catalogue', [1853].
85. Sullivan, *British Literary Magazines*, 505.
86. Hoag, 'Shipboard Printing', 102.
87. Max Keith Sutton, '"Inverse Sublimity" in Victorian Humour', *Victorian Studies* 10, no. 2 (December 1966): 177–92 at p. 188.
88. Ann C. Colley, *Victorians in the Mountains: Sinking the Sublime* (London: Routledge, 2016), 39–42.
89. Sutton, 'Victorian Humour', 182.
90. Mary Douglas, 'The Social Control of Cognition: Some Factors in Joke Perception', *Man* 3, no. 3 (September 1968): 361–76 at pp. 363; 368.
91. Altick, *Punch*, p. xxiv.
92. Aimar Ventsel, '"Our Beer is Better Than Theirs": Sub-cultural Comics as a Joking Relationship with Identity', *Folklore: Electronic Journal of Folklore*, no. 46 (2010): 143–60 at p. 143.
93. Ibid., 156.
94. Colley, *Victorians in the Mountains*, 30–2.
95. Osborn, 'Queen's Illuminated Magazine', 4 [February 1853]: 7–10.
96. Ibid., 10.
97. George Back, *Narrative of an Expedition in H.M.S. Terror* (London: John Murray, 1838), 427.
98. Osborn, 'Queen's Illuminated Magazine', 4 [February 1853]: 8. The missing word has the same number of letters as the *Terror*.
99. Ibid., 4 [February 1853]: 9. Samuel Taylor Coleridge's 'Rime of the Ancient Mariner' influenced literary depictions of ice throughout the nineteenth

century. The use of quotations from the poem in Arctic narratives is discussed in Chapter 3.

100. Ibid., 4 [February 1853]: 9.
101. Ibid., 4 [February 1853]: 10.
102. Owen Beattie and John Geiger, *Frozen in Time: The Fate of the Franklin Expedition* (London: Bloomsbury, 1987).
103. Osborn, 'Queen's Illuminated Magazine', 1 [November 1852]: 8.
104. It was felt by writers such as William Gilpin that cows, with their rough coats, fitted with the picturesque aesthetic. Gilpin believed that three cows was the ideal number to have in a picture. William Gilpin, *Observations, on Several Parts of England, Particularly the Mountains and Lakes of Cumberland and Westmoreland* (London: Cadell and Davies, 1808), xl–xli.
105. Osborn, 'Queen's Illuminated Magazine', 4 [February 1853]: 13. Cape Farewell is at the southern tip of Greenland and would be passed on the return journey to Britain.
106. Osborn, 'Queen's Illuminated Magazine', 2 [December 1852]: 20.
107. Davis-Fisch, *Loss and Cultural Remains*, 29–30.
108. Osborn and McDougall, *Illustrated Arctic News*, 17.
109. *Arctic Miscellanies*, 41.
110. Eveline G. Bouwers, *Public Pantheons in Revolutionary Europe: Comparing Cultures of Remembrance, c. 1790–1840* (New York: Palgrave Macmillan, 2012), 202–7.
111. Abraham Bradford, Journal of Her Majesty's Sloop Resolute [28 February 1850–9 October 1851], ADM 101/117/3, TNA UK; Osborn and McDougall, *Illustrated Arctic News*, 44.
112. Stam and Stam, 'Bending Time', 302; Behrisch Elce, 'Bright Objects', 347.
113. Behrisch Elce, 'Bright Objects', 344.
114. Johann Miertsching, *Frozen Ships: The Arctic Diary of Johann Miertsching 1850–1854*, trans. L. H. Neatby (Toronto: Macmillan, 1967), 176.
115. J. D. Gilpin, 'Outline of the Voyage of H.M.S. Enterprize and Investigator to Barrow Strait in search of Sir John Franklin', *Nautical Magazine* 19 (January 1850), 8–19, 82–90, 160–70, 230.
116. Quoted in Markham, *Sir Clements R. Markham*, 120.
117. For example, those involved in the production of an Arctic periodical could instantly see and hear the reactions of the entire readership (the ship's company) in a way that was not possible in the metropole.
118. Two nights of theatre and a masquerade, followed by late-night singing and dancing, were held during September 1849 on the *Herald* and the *Plover*. William Simpson, Diary Written on board HMS Plover [January 1848–December 1850], JOD/76: Entry for 18 and 29 September 1849, NMM, Greenwich; [Chimmo], *Euryalus*, 299–311.

119. Simpson, 'Flight of the Plover', [1848].
120. Clements Markham, *Life of Sir Admiral Leopold McClintock* (London: John Murray, 1909), 113.
121. Markham, *Sir Clements R. Markham*, 120. As neither of the two periodicals apparently survives, only second-hand references to them are available.
122. Blum, 'Polar Periodicals', 163.
123. The present-day communities at Resolute (Qausuittuq) and Grise Fiord (Aujuittuq) were created in the 1950s by the forced relocation of Inuit from other areas. Arctic Bay, a distance of 350 km to the southeast as the crow flies, would have been the closest Inuit settlement.
124. Sherard Osborn, *Stray Leaves from an Arctic Journal* (London: Longman, Brown, Green, and Longmans, 1852), 192.
125. *Arctic Miscellanies*, 275–7; 264–7; 284.
126. Osborn, 'Queen's Illuminated Magazine', Supplement [January 1853]: 6. The supplement was part of the January issue and is bound with the entire manuscript, because so many contributions had been received from expedition members.
127. Two titles by G. P. R. James, *The Robber* (1838) and *The King's Highway* (1840), were listed in the library catalogue of the *Assistance* in 1853.
128. By 1852, when the expedition left England, Frances Milton Trollope had published numerous works. Her son, Anthony Trollope, to whom the allusion could also refer, also became a popular novelist. Although the latter had published three novels by 1852, it was not until the late 1850s that he became well known. *Literature Online*, s.v. 'Trollope, Anthony', by Mark Turner (Cambridge: Chadwyck-Healey, 2001).
129. Rochfort Maguire, *The Journal of Rochfort Maguire 1852–1854*, ed. John Bockstoce (London: The Hakluyt Society, 1988), 1:85.
130. William Hulme Hooper, *Ten Months among the Tents of the Tuski, with Incidents of an Arctic Boat Expedition in Search of Sir John Franklin, as far as the Mackenzie River, and Cape Bathurst* (London: John Murray, 1853); Maguire, *Journal*, 1988.
131. Miertsching, *Frozen Ships*, 36–7; Maguire, *Journal*, 1:130–1; McGoogan, *Fatal Passage*, 126–7.
132. Hoag, 'Shipboard Printing', 40–2.
133. Matthews, Journal, Entry for February 1853.
134. Blum, 'Polar Periodicals', 176–7.
135. Blum, 'Polar Periodicals', 173–6.
136. *Arctic Miscellanies*, p. viii.
137. Colburn & Co. Publishers, handbill advertising *Arctic Miscellanies*, [1852]. The price was twelve shillings.
138. 'Literature', *Era*, 3 May 1852.

139. 'The Fac-simile of "The Illustrated Arctic News", Published On Board H. M.S. Resolute', *Literary Gazette*, 10 April 1852.
140. 'Arctic Miscellanies', *Lloyd's Weekly Newspaper*, 4 January 1852.
141. 'Illustrated Arctic News', *Literary Gazette*, 10 April 1852.
142. Osborn, 'Queen's Illuminated Magazine'. Osborn's note also confirms that the paper was made up of contributions and not merely written by him.
143. Ibid.
144. May, Journal, Entry for 11 October 1852.
145. I am grateful to Peter Martin for drawing my attention to the lack of marginalia, in response to a paper I gave on the 'Hidden Histories of Polar Exploration' panel at the International Conference of Historical Geographers, 2018.
146. It seems that initially the paper was called the 'Victoria Gazette' but that at some point before the production of the first issue its name was changed to the 'Queen's Illuminated Magazine'.
147. May, Journal, Entry for 11 October 1852.
148. *Arctic Miscellanies*, 248.
149. Other types of souvenirs were created during the voyage by the carpenter William T. Mumford, on Kellett's arm of the Belcher expedition, who 'completed the fleet of model sledges to the number of eleven'. William T. Mumford, Private Journal of an Expedition to the Arctic Regions to Ascertain the Fate of Sir John Franklin ... in the Years 1852. 3. 4 [1 March 1852–24 October 1854], W. T. Mumford Fonds [1852–97], MG24-H80: Entry for 11 February 1854, LAC, Ottawa; James Nelson to William T. Mumford, 9 September [Year?], in Mumford, Private Journal.
150. Amanda Gilroy, 'Introduction', in *Romantic Geographies: Discourses of Travel 1775–1844*, ed. Amanda Gilroy (Manchester: Manchester University Press, 2000), 1–15 at p. 3.
151. Osborn, 'Queen's Illuminated Magazine', 4 [February 1853]: 28.
152. Osborn, 'Queen's Illuminated Magazine', 3 [January 1853]: 24.
153. Osborn, 'Queen's Illuminated Magazine', 4 [February 1853]: 28.
154. It is notable that, while expedition members used the word 'sojourn' to refer to their winter in the ice, popular media was fond of using the word 'imprisonment'.

3 'These Dread Shores'

1. Hester Lynch Piozzi to John Williams, 12 April 1819, in *Electronic Enlightenment Scholarly Edition of Correspondence*, ed. Robert McNamee et al. (Oxford: University of Oxford, 2008–17), www.e-enlightenment.com. The book she referred to was the

narrative of John Ross, *Voyage of Discovery . . . for the Purpose of Exploring Baffin's Bay* (London: John Murray, 1819).

2. That number does not include three folios of lithographs that included short written accounts and could arguably be thought of as pictorial narratives; two published versions of on-board periodicals; subsequent editions of several books; journals published posthumously after the period; or compilations put together by publishers.
3. Julia Thomas, *Pictorial Victorians: The Inscription of Values in Word and Image* (Athens: Ohio University Press, 2004), 8.
4. As far as we know, all the authors were male.
5. E. A. Inglefield, *A Summer Search for Sir John Franklin with a Peep into the Polar Basin* (London: Thomas Harrison, 1853), 41.
6. Mary Louise Pratt, *Imperial Eyes: Travel Writing and Transculturation* (London: Routledge, 1992), 201.
7. Pratt, *Imperial Eyes*, 61.
8. Tim Cresswell, *Place: A Short Introduction* (Oxford: Blackwell, 2004), 7.
9. Yi-Fu Tuan, *Space and Place: The Perspective of Experience* (Minneapolis: University of Minnesota Press, 1977), 6.
10. Kirsten Hvenegård-Lassen, 'Journey to the Centre of the Ice: Narrating Ice-Core Drillings in Northern Greenland', *Studies in Travel Writing* 20, no. 3 (September 2016): 237–48 at p. 237.
11. Sheila Watt-Cloutier, *The Right to Be Cold: One Woman's Fight to Protect the Arctic and Save the Planet from Climate Change* (Minneapolis: University of Minnesota Press, 2018), p. xxv.
12. Jen Hill, *White Horizon: The Arctic in the Nineteenth-Century British Imagination* (Albany: State University of New York Press, 2008), 42.
13. Nanna Katrine Lüders Kaalund, 'What Happened to John Franklin? Danish and British Perspectives from Francis McClintock's Arctic Expedition, 1857–59', *Journal of Victorian Culture* 25, no. 2 (2020): 300–14 at p. 311.
14. Adriana Craciun, 'Writing the Disaster: Franklin and *Frankenstein*', *Nineteenth-Century Literature* 65, no. 4 (March 2011): 433–80 at pp. 435–7.
15. *The English Catalogue of Books from 1835 to 1863* (London: Sampson Low, 1864). Clements Markham, *Franklin's Footsteps: A Sketch of Greenland, along the Shores of which His Expedition Passed, and of the Parry Isles, where the Last Traces of It were Found* (London: Chapman and Hall, 1853); Edward Belcher, *The Last of the Arctic Voyages* (London: Lovell Reeve, 1855).
16. Craciun, 'Franklin and *Frankenstein*', 457.
17. Sherard Osborn, *Stray Leaves from an Arctic Journal* (London: Longman, Brown, Green, and Longmans, 1852).

18. Innes M. Keighren, Charles W. J. Withers, and Bill Bell, *Travels into Print: Exploration, Writing, and Publishing with John Murray, 1773–1859* (Chicago and London: The University of Chicago Press, 2015), 1.
19. Barbara Korte, *English Travel Writing from Pilgrimages to Postcolonial Explorations*, trans. Catherine Matthias (Houndmills, Basingstoke: Macmillan, 2000), 88.
20. See Adriana Craciun, 'What is an Explorer?', *Eighteenth-Century Studies* 45, no. 1 (Fall 2011): 29–51.
21. Henry Briant, 'A Catalogue of the Library Established on board HMS Assistance' [1853], Arctic Pamphlets vol. 3, mg N07/08K, RGS, London. As well as the library, officers tended to bring their own books on board.
22. R. H. Dana, *Two Years before the Mast: A Personal Narrative of Life at Sea*, 4th ed. (London: Edward Moxon, 1845).
23. The haphazard nature of the library catalogue, printed on board by the clerk, makes exact calculations impossible. For example, entries may be listed only by author or by one word of the title or in the following manner: 'Mast. Two Years before the'. Briant, 'Catalogue', 17.
24. For more on the literary technique of virtual witnessing, see Steven Shapin and Simon Schaffer, *Leviathan and the Air-Pump: Hobbes, Boyle, and the Experimental Life* (Princeton: Princeton University Press, 1989), 60–5.
25. Marina Warner, 'Imaginary Passages', *Third Text* 21, no. 3 (September 2007): 487–98 at p. 488.
26. Osborn, *Stray Leaves*, 157.
27. Osborn, *Stray Leaves*, 156.
28. Ibid.
29. Ibid., 157.
30. Ibid., 157–8.
31. William Hulme Hooper, *Ten Months among the Tents of the Tuski, with Incidents of an Arctic Boat Expedition in Search of Sir John Franklin, as far as the Mackenzie River, and Cape Bathurst* (London: John Murray, 1853), 408–9.
32. Ibid., 409–10.
33. Ibid., 412.
34. Isaac Israel Hayes, *An Arctic Boat Journey, in the Autumn of 1854*, 2nd ed. (Boston: Ticknor and Fields, 1867), 250–1.
35. William Parker Snow, *Voyage of the Prince Albert in Search of Sir John Franklin: A Narrative of Every-Day Life in the Arctic Seas* (London: Longman, Brown, Green, and Longmans, 1851), 249.

36. Felix Driver and Luciana Martins, 'Views and Visions of the Tropical World', in *Tropical Visions in an Age of Empire*, ed. Felix Driver and Luciana Martins (Chicago: University of Chicago Press, 2005), 3–20 at p. 8.
37. Alexander von Humboldt, 'Landscape Painting', in *Cosmos: Sketch of a Physical Description of the Universe*, vol. 2, trans. under the superintendence of Edward Sabine (London: Longman, Brown, Green, and Longmans and John Murray, 1849), 90–1.
38. Osborn, *Stray Leaves*, 35.
39. Hooper, *Tents of the Tuski*, 355.
40. Carl Thompson, *The Suffering Traveller and the Romantic Imagination* (Oxford: Clarendon Press, 2007), 62.
41. Bernard Smith, *Imagining the Pacific: In the Wake of the Cook Voyages* (New Haven: Yale University Press, 1992), 135–40.
42. Craciun, 'Franklin and *Frankenstein*', 449; Ian S. MacLaren, 'The Aesthetic Map of the North, 1845–1859', *Arctic* 28, no. 2 (June 1985): 89–103; Susan Birkwood, 'From "Naked Country" to "Sheltering Ice": Rudy Wiebe's Revisionist Treatment of John Franklin's First Arctic Narrative', *Nordlit* 23 (2008): 25–38.
43. Osborn, *Stray Leaves*, 70.
44. Ann Radcliffe, *A Journey Made in the Summer of 1794, through Holland and the Western Frontier of Germany*, 2 vols. (London: G. G. and J. Robinson, 1795) 1: p. vi.
45. During the late 1840s, many of the officers who later served on the Arctic search expeditions, including Sherard Osborn, William Henry Browne, Frederick George Mecham, and Francis Leopold McClintock, were serving on naval ships, some on survey missions, in the Pacific.
46. Hooper, *Tents of the Tuski*, 29.
47. Osborn, *Stray Leaves*, 116. Tahiti was known as Otaheite in this period.
48. Heidi Hansson, 'The Gentleman's North: Lord Dufferin and the Beginnings of Arctic Tourism', *Studies in Travel Writing* 13, no. 1 (February 2009): 61–73 at p. 64.
49. 'Calving' occurs when chunks of ice break off a glacier's terminus.
50. Robert Goodsir, *An Arctic Voyage to Baffin's Bay and Lancaster Sound, in Search of Friends with Sir John Franklin* (London: John van Voorst, 1850), 53.
51. Marina Warner, *Once Upon a Time: A Short History of Fairy Tale* (Oxford: Oxford University Press, 2014), 24.
52. A schooner is 'a small craft with two masts'. W. H. Smyth, *An Alphabetical Digest of Nautical Terms*, ed. Edward Belcher (London: Blackie and Son, 1867).
53. Inglefield, *Summer Search*, 14.

54. Inglefield, *Summer Search*, 41. The phrase 'grim watchmen' is likely taken from a poem, 'Queen Victoria at Windsor' by John Walker Ord, and refers to 'giant oaks': 'These stalwart boughs for tempests have no fear, / But seem as steadfast *now* as in their prime: / Grim watchmen of the past–in hoary age sublime!' *The Bard, and Minor Poems*, ed. John Lodge, 2nd ed. (London: Simpkin, Marshall. & Co., 1842), 159–64.
55. See for example Hayes, *Arctic Boat Journey*, 285–6.
56. Ann Radcliffe, *The Mysteries of Udolpho: A Romance; Interspersed with Some Pieces of Poetry*, 4 vols., 4th ed. (London: G. G. and J. Robinson, 1799), 2:170–1.
57. 'As may well be imagined my time was now fully engaged, and my pencil and sextant were rarely out of my hand by night more than day.' Inglefield, *Summer Search*, 66.
58. Cian Duffy, *The Landscapes of the Sublime, 1700–1830: Classic Ground* (Houndmills, Basingstoke: Palgrave Macmillan, 2000), 105.
59. Inglefield, *Summer Search*, 76; 115. All of the plates in the book can be viewed in high resolution on the website *The Illustration Archive*: https://illustratio narchive.cf.ac.uk.
60. Inglefield, *Summer Search*, 107; 8.
61. Ibid., 65. The theory of the Open Polar Sea gained currency in the late eighteenth century and became popular during the Franklin searches. Its proponents believed that the North Pole was situated in an open sea surrounded on the fringes by ice.
62. Craciun, 'Franklin and *Frankenstein*', 463.
63. Ibid., 435.
64. Andrew Lambert, *Franklin: Tragic Hero of Polar Navigation* (London: Faber & Faber, 2009), 200.
65. Janice Cavell, *Tracing the Connected Narrative: Arctic Exploration in British Print Culture 1818–1860* (Toronto: University of Toronto Press, 2008), 227.
66. Hayes, *Arctic Boat Journey*, 186. Several new and illustrated editions of Hayes's book were published in the 1860s and 1870s, including a French translation.
67. James Beattie, *The Minstrel; or, the Progress of Genius*, bk. 1 (London: Dilly, 1771), 31.
68. Robert Southey, *The Life of Nelson* (London: Collins, 1813), 24.
69. Tim Fulford, 'Romanticism and Colonialism: Races, Places, Peoples, 1800–30', in *Romanticism and Colonialism: Writing and Empire, 1789–1830*, ed. Tim Fulford and Peter J. Kitson (Cambridge: Cambridge University Press, 1998), 35–47 at p. 37.
70. Constantine Phipps, *A Voyage towards the North Pole* (London: Bowyer and Nichols, 1774).

71. Many men participated in successive voyages and spent several winters in the Arctic. Some of the crew of the *Plover* voluntarily spent six winters, from 1848 to 1854, in the Bering Strait area.
72. For example: Officers and Seamen of the Expedition, *Arctic Miscellanies: A Souvenir of the Late Polar Search*, ed. James Donnet (London: Colburn & Co., 1852), 248; 261.
73. Korte, *English Travel Writing*, 105.
74. 1848; 1850; 1852; 1853; 1857; 1858.
75. Francis McClintock, *The Voyage of the 'Fox' in the Arctic Seas: A Narrative of the Discovery of the Fate of Sir John Franklin and His Companions* (London: John Murray, 1859), p. iii; 17; 26.
76. MacLaren, 'Aesthetic Map', 95.
77. George Frederick McDougall, *The Eventful Voyage of H. M. Discovery Ship 'Resolute' to the Arctic Regions* (London: Longman, Brown, Green, Longmans, & Roberts, 1857), 95.
78. Elisha Kent Kane, *The U.S. Grinnell Expedition in Search of Sir John Franklin. A Personal Narrative* (New York: Harper & Brothers, 1854), 43; 44; 46–7.
79. Ibid., 46–7.
80. Kane, *U.S. Grinnell Expedition*, p. v.
81. Peter Sutherland, *Journal of a Voyage in Baffin's Bay and Barrow Straits . . . in Search of the Missing Crews of H.M. Ships Erebus and Terror*(London: Longman, Brown, Green, and Longmans, 1852) 2: pl. 4.
82. By contrast, Snow's narrative, *Voyage of the Prince Albert,* was priced at twelve shillings.
83. Sutherland, *Journal of a Voyage*, p. xliii.
84. Sutherland, *Journal of a Voyage*, p. xliii.
85. McClintock, *Voyage of the 'Fox'*, 152.
86. Osborn, *Stray Leaves*, 47.
87. Ann Radcliffe, 'The Snow-Fiend', in *Gaston de Blondeville and St. Alban's Abbey*, 4 vols. (London: Henry Colburn, 1826), 4:192–5. The expeditions were much further north than Newfoundland and Labrador.
88. Sadiah Qureshi, *Peoples on Parade: Exhibition, Empire, and Anthropology in Nineteenth-Century Britain* (Chicago: University of Chicago Press, 2011), 88; 177.
89. Efram Sera-Shriar, 'Arctic Observers: Richard King, Monogenism and the Historicisation of Inuit through Travel Narratives', *Studies in History and Philosophy of Biological and Biomedical Sciences* 51 (2015): 23–31 at p. 24.
90. Ibid., 6. Anthropology became popular in 1860s as a broader method of study.
91. James Cowles Prichard, 'Ethnology', in *A Manual of Scientific Enquiry: Prepared for the Use of Her Majesty's Navy and Adapted for Travellers in*

General. Published by Authority of the Lords Commissioners of the Admiralty, ed. John F. W. Herschel (London: John Murray, 1849), 423–44.

92. Ibid., 423–4.
93. Sera-Shriar, 'Arctic Observers', 23.
94. Ellen Boucher, 'Arctic Mysteries and Imperial Ambitions: The Hunt for Sir John Franklin and the Victorian Culture of Survival', *The Journal of Modern History* 90 (March 2018): 40–75 at p. 61.
95. Qureshi, *Peoples on Parade*, 168.
96. Pratt, *Imperial Eyes*, 4.
97. McClintock, *Voyage of the 'Fox'*, 27.
98. McDougall, *Eventful Voyage*, 72.
99. Qureshi, *Peoples on Parade*, 214–15.
100. Prichard, 'Ethnology', 430; 426. Explorers were not averse to digging up Inuit graves.
101. Goodsir, *Arctic Voyage*, 27.
102. Pratt, *Imperial Eyes*, 63.
103. Efram Sera-Shriar, *The Making of British Anthropology, 1813–1871* (London: Pickering and Chatto, 2013), 140.
104. Inglefield, *Summer Search*, 128.
105. Janice Cavell, review of *Writing Arctic Disaster: Authorship and Exploration*, by Adriana Craciun, *Journal of Historical Geography*. Published electronically September 25, 2016: 2.
106. Private, unpublished journals from the Bering Strait region also attest to social interaction in the contact zone. For example, when three Iñupiat men died after being driven out to sea on a piece of ice, John Matthews, the *Plover*'s boatswain, recorded their loss with regret: 'One of them was a great friend of mine, the Whale Chief.' John Matthews, The Journal of John Matthews from September 1850 to April 1855, written during a voyage in search of the Franklin Expedition, [1850–5] mg N07/11D: Entry for January 1854. Monograph Collection, RGS, London.
107. Hooper, *Tents of the Tuski*, 1853. All of the plates in the book can be viewed in high resolution on the website *The Illustration Archive*: https://illustrationarchive.cf.ac.uk. The entire book with colour plates can be viewed on the TPL website: https://static.torontopubliclibrary.ca/da/pdfs/37131055418750d.pdf.
108. Hooper, *Tents of the Tuski*, 162.
109. It is not known who made the initial sketches in the Arctic; those in the book are attributed to an F. Skill. There was no F. Skill listed on the ship's muster list. The captions on most of the illustrations are short, further suggesting that the artist was not on the expedition.
110. Hooper, *Tents of the Tuski*, 71–2.

111. Uvedale Price, *An Essay on the Picturesque, as Compared with the Sublime and the Beautiful; and, on the Use of Studying Pictures, for the Purpose of Improving Real Landscape* (London: Robson, 1796), 59.
112. Hooper, *Tents of the Tuski*, 71.
113. Pratt, *Imperial Eyes*, 61.
114. Hooper, *Tents of the Tuski*, 306.
115. Hooper, *Tents of the Tuski*, 211.
116. Some of the work by Aron from Kangeq in the Ethnographic Museum of the University of Oslo from 1860 shares similar traits, for example, *Tuluit Nunaliartut* [1860], UEM30049, Kulturhistorisk museum, Oslo.
117. Goodsir, *Arctic Voyage*, 33.
118. McDougall, *Eventful Voyage*, 74–5.
119. *Oxford English Dictionary*: 'Of persons: Fair, pretty, "nice-looking", pleasing: in modern use implying a lower or homelier style of personal beauty, which pleases but does not excite admiration.' 'Hideous' on the other hand, implies something 'repulsive, revolting'.
120. Hooper, *Tents of the Tuski*, 51.
121. Ibid., 51.
122. Ibid, 73.
123. Ibid., 206.
124. Ibid., 28.
125. Korte, *English Travel Writing*, 92.
126. Osborn, *Stray Leaves*, 155.
127. Goodsir, *Arctic Voyage*, 4.
128. Goodsir, *Arctic Voyage*, 13.
129. Osborn, *Stray Leaves*, 207. I deal with the paucity of printed representations of sledge dogs in Chapter 5.
130. Osborn, *Stray Leaves*, 204. In this case, the text of the narrative concerning the men is more positive than the scene transferred to print.
131. For example, Walter May prepared some drawings for McClintock's narrative, despite not having been on the expedition, and a later edition of Hayes's *Arctic Boat Journey* included illustrations of selected passages of the narrative text, obviously produced long after Hayes's experience.
132. Qureshi, *Peoples on Parade*, 94–5.
133. Keighren, Withers, and Bell, *Travels into Print*, 158.
134. Ibid., 157.
135. Kane, *U.S. Grinnell Expedition*, 1854.
136. The entire book with colour images is available online: https://static.torontopubliclibrary.ca/da/pdfs/37131055492003d.pdf.
137. Keighren, Withers, and Bell, *Travels into Print*, 155.
138. Osborn, *Stray Leaves*, p. vi.

139. Christine E. Jackson, 'M & N Hanhart: Printers of Victorian Music Covers', *Fontes Artis Musicae* 39, no. 3/4 (July–December 1992): 326–30 at p. 326. In the mid-nineteenth century, sheet music with illustrated covers, for popular songs intended to be performed in the home, could sell as many as 200,000 copies. Jackson, 'M & N Hanhart', 327.
140. Osborn, *Stray Leaves*, p. v.
141. Osborn, *Stray Leaves*, 129.
142. Sherard Osborn and George F. McDougall, *Illustrated Arctic News* (London: Ackermann, 1852), 20. Each page of the *Illustrated Arctic News* can be viewed online on the Royal Museums Greenwich website: https://collections.rmg.co.uk/collections/objects/137953.html.
143. Osborn, *Stray Leaves*, 128.
144. The graves were those of three members of the Franklin Expedition who had died during the first year of the voyage.
145. MacDougall gives a figure of 334 expedition members in his narrative. *Eventful Voyage*, p. xxxviii. My own calculation using the muster lists is 303. Neville W. Poulsom, *British Polar Exploration and Research: A Historical and Medallic Record with Biographies 1818–1999* (London: Savannah Publications, 2000), 64–7; 87–9; 91–2.
146. McDougall, *Eventful Voyage*, 79; Osborn, *Stray Leaves*, 107–20.
147. Osborn and McDougall, *Illustrated Arctic News*, 2.
148. George McDougall, *The 3 Graves – Beechey Is.d* [1850], PAG7997, NMM, Greenwich.
149. The letters 'A.B.' stand for Able Seaman, 'one who fully knew his trade'. This is above the Ordinary Seaman who was 'younger and less experienced'. Marcus Rediker, *Between the Devil and the Deep Blue Sea*, Canto edition (Cambridge: Cambridge University Press, 1993), 85.
150. Kane, *U.S. Grinnell Expedition*, 162; Lambert, *Franklin: Tragic Hero*, 201.
151. Mark Metzler Sawin, 'Raising Kane: Elisha Kent Kane and the Culture of Fame in Antebellum America', *Transactions of the American Philosophical Society* 98, no. 3 (2008): 1–368 at p. 21.
152. Canadian Coast Guard, 'Ice Climatology and Environmental Conditions', in *Ice Navigation in Canadian Waters* (Ottawa: Government of Canada, 2014), www.ccg-gcc.gc.ca/publications/icebreaking-deglacage/ice-navigation-glaces/docs/ice-navigation-dans-les-galces-eng.pdf.
153. Kane, *U.S. Grinnell Expedition*, 162.
154. Lambert, *Franklin: Tragic Hero*, 202.
155. Kane, *U.S. Grinnell Expedition*, 85; 95.
156. William Henry Browne, *The Devil's Thumb, Ships Boring and Warping in the Pack* (London: Ackermann, 1850). Browne's lithographs are discussed in more detail in Chapter 5. The set of lithographs can be viewed on the Royal

Collection Trust website: www.rct.uk/collection/search#/1/collection/750930/arctic-expedition-in-search-of-sir-john-franklin-1848–49.

157. Jules Verne, *The Adventures of Captain Hatteras*, trans. unknown (London: Ward Lock & Co., 1876), 93–8.
158. McClintock, *Voyage of the 'Fox'*, 3rd ed., 1869. The first edition contained sixteen illustrations and four maps. The third edition had thirty illustrations and three maps.
159. McClintock, *Voyage of the 'Fox'*, 71.
160. Tabular icebergs have near vertical sides and flat tops.
161. McClintock, *Voyage of the 'Fox'*, 3rd ed., 293.
162. Kane, *U.S. Grinnell Expedition*, 79.
163. McClintock, *Voyage of the 'Fox,'* 146–7.
164. Goodsir, *Arctic Voyage*, 23.
165. Inglefield, *Summer Search*, 14.
166. Pratt, *Imperial Eyes*, 209.

4 'Never to Be Forgotten'

1. Tom Taylor, *Leicester Square; Its Associations and its Worthies* (London: Bickers and Son, 1875), 467.
2. Ralph Hyde, *Panoramania! The Art and Entertainment of the 'All-Embracing' View* (London: Trefoil, 1988), 13. A valuable resource for the study of both moving and fixed (stable) panoramas is Laurie Garrison, ed., *Panoramas, 1787–1900: Texts and Contexts*, 5 vols. (London: Pickering and Chatto, 2013).
3. Hyde, *Panoramania*, 39. To put this cost into perspective, lotions, powders, and remedies such as 'Johnson's American Soothing Syrup', for the relief of teething in children, were advertised at over twice the price of entry to a panorama. Classified Advertisements, *Observer*, 20 January 1850.
4. 'The Drama and Public Amusements', *Critic*, 1 December 1849.
5. Like the majority of panoramas from the nineteenth century, *Summer and Winter Views of the Polar Regions* does not survive. Sometimes panoramas were sent to other cities to be exhibited; some panoramas were destroyed by fire; storage of these immense paintings was problematic, and many panoramas were 'obliterated' by painting a new scene on the canvas. 'Panoramas', *Chambers's Journal of Popular Literature, Science and Arts* 316, 21 January 1860.
6. William Henry Browne, letter to the editor, *The Times*, 29 December 1849; *Literary Gazette*, 5 January 1850; *Athenaeum*, 12 January 1850. Browne further stated in the letter that he was 'the only officer or person in the Enterprise who took any drawings of those regions during the late expedition under Sir

James Ross, and which drawings the Admiralty have allowed Mr. Burford, of Leicester Square, to use'.

7. 'Panorama of the Polar Regions', *Literary Gazette*, 18 February 1850; 'Burford's Panorama of the Polar Regions', *Illustrated London News*, 23 February 1850.
8. 'The Drama and Public Amusements', *Critic*, 15 November 1850.
9. Earlier Arctic exploration had inspired panoramas such as *North Pole* (1819) and *Spitzbergen* (1820) in London and *Frozen Regions* (1820) in Dublin. Space does not permit an exploration of these fascinating panoramas here; Kevin Rockett and Emer Rockett give an excellent account of the Arctic panorama in Dublin, while Russell Potter provides an exceptional survey of the London panoramas. Kevin Rockett and Emer Rockett, *Magic Lantern, Panorama and Moving Picture Shows in Ireland, 1786–1909* (Dublin: Four Courts Press, 2011); Russell Potter, *Arctic Spectacles: The Frozen North in Visual Culture 1818–1875* (Seattle and London: University of Washington Press, 2007).
10. *Lloyd's Weekly Newspaper*, 18 November 1849.
11. From advertisements, it is not entirely clear what form *View of the Polar Regions* took, but the *Examiner* noted that the view was 'very graphically painted, on a small scale, by Messers Danson and Son'. It was accompanied by the *Ruins of Netley Abbey* and *Tchin Shau, or Golden Island*, in China. 'Miscellaneous', *Examiner*, 29 December 1849; Advertisement, *Examiner*, 12 January 1850; Advertisement, *Observer*, 20 January 1850.
12. 'Christmas Exhibitions', *Theatrical Journal*, 3 January 1850. A moving panorama involved a long, horizontal canvas attached to rollers. The painting was unrolled to display a sequence of scenes before an audience and was often accompanied by music or commentary. From 19 July to 21 August 1850, this panorama was showing in Dublin twice a day, along with 'appropriate music and descriptive lecture and anecdotes'. A 'descriptive book' of the panorama was also available. *Freeman's Journal*, 19 July 1850.
13. Advertisement, *Athenaeum*, 30 March 1850; Advertisement, *Athenaeum*, 4 May 1850.
14. Laurie Garrison, 'Virtual Reality and Subjective Responses: Narrating the Search for the Franklin Expedition through Robert Burford's Panorama', *Early Popular Visual Culture* 10, no. 1 (February 2012): 7–22 at p. 20.
15. Ibid., 7.
16. Ibid., 17. The Ross expedition was deemed a failure by the press, having returned with no trace of Franklin. Janice Cavell, *Tracing the Connected*

Narrative: Arctic Exploration in British Print Culture 1818–1860 (Toronto: University of Toronto Press, 2008), 183–4.

17. Garrison, 'Virtual Reality', 14–15.
18. Ibid., 19–20.
19. Potter, *Arctic Spectacles*, 82.
20. Ibid., 83
21. Robert G. David, *The Arctic in the British Imagination 1818–1914* (Manchester: Manchester University Press, 2000), 154.
22. Garrison, 'Virtual Reality', 8; Ralph Hyde, 'Burford, Robert (1791–1861)', *Oxford Dictionary of National Biography* (Oxford: Oxford University Press, 2004), www.oxforddnb.com/view/article/3965?docPos=2.
23. Burford, 'Summer and Winter Views', 10.
24. *Athenaeum*, 29 December 1849.
25. Sadiah Qureshi, *Peoples on Parade: Exhibition, Empire, and Anthropology in Nineteenth-Century Britain* (Chicago: University of Chicago Press, 2011), 7–8.
26. 'Panoramas', *Chambers's Journal of Popular Literature, Science and Arts* 316, 21 January 1860.
27. Richard D. Altick, *The Shows of London* (Cambridge, MA: Belknap Press of Harvard University Press, 1978), 137–8.
28. Ibid., 137.
29. Burford, 'Summer and Winter Views', 10. Distemper was a cheaper, non-durable, water-based medium.
30. John Ruskin, 'Deucalion, and Other Studies in Rocks and Stones' (1878), in *Library Edition of the Works of John Ruskin*, ed. E. T. Cook and Alexander Wedderburn, 39 vols. (London: George Allen, 1903–12), 26:567. www.lancaster.ac.uk/users/ruskinlib/Pages/Works.html
31. John Ruskin, 'Præterita. Outlines of Scenes and Thoughts Perhaps Worthy of Memory in my Past Life' (1886–9), in Cook and Wedderburn, *Works of John Ruskin*, 35:118.
32. Taylor, *Leicester Square*, 467.
33. 'Miscellaneous', *Musical World*, 29 June 1850.
34. Taylor, *Leicester Square*, 469.
35. Rockett and Rockett, *Magic Lantern*, 91–2.
36. Ruskin, 'Præterita', in Cook and Wedderburn, *Works of John Ruskin*, 35:118.
37. Alexander von Humboldt, 'Landscape Painting', in *Cosmos: Sketch of a Physical Description of the Universe*, vol. 2, trans. under the superintendence of Edward Sabine (London: Longman, Brown, Green, and Longmans and John Murray, 1849), 91.
38. Ibid., 90.
39. *A Section of the Rotunda in Leicester Square*. From Robert Mitchell, *Plans and Views in Perspective of Buildings Erected in England and Scotland; and . . . an*

essay to elucidate the Grecian, Roman and Gothic Architecture (London: Wilson and Co., 1801), plate 14. This section shows a view before the third storey was added.

40. I am grateful to Elizabeth Tilley for drawing my attention to this detail regarding the women. The print shows twenty-three women, ten men, and two children visiting the panorama. A high-resolution digital version of this image is available online from the British Library Collections: https://www.bl.uk/collection-items/section-of-the-rotunda-leicester-square#.
41. 'Panoramas', *Chambers's Journal of Popular Literature, Science and Arts* 316, 21 January 1860.
42. Altick, *Shows of London*, 133.
43. Humboldt, 'Landscape Painting', 90.
44. 'Miscellaneous', *Examiner*, 29 December 1849.
45. 'Madame Tussaud's', *Theatrical Journal*, 3 January 1850.
46. The average length of a panorama's run at Leicester Square during the late 1840s and 50s was about a year. Altick, *Shows of London*, 140.
47. 'Panorama of the Arctic Regions', *Observer*, 10 February 1850.
48. 'The Drama and Public Amusements', *Critic*, 15 April 1850.
49. 'Drama and Public Amusements', *Critic*, 15 November 1850.
50. Arctic panoramas had twice been painted at this venue. In 1819, *View of the North Coast of Spitzbergen* had been based on the drawings of Lieutenant Beechey, and in 1834, Burford and Selous painted *A View of the Continent of Boothia*, based on Captain John Ross's visual records.
51. Burford, 'Summer and Winter Views', 15; 'Mr. Burford's Panorama', *The Times*, 11 February 1850.
52. Humboldt, 'Landscape Painting', 90–1.
53. William Parker Snow, *Voyage of the Prince Albert in Search of Sir John Franklin: A Narrative of Every-Day Life in the Arctic Seas* (London: Longman, Brown, Green, and Longmans, 1851), 249.
54. 'Professor Leslie's Lectures on Painting', *Athenaeum*, 17 February 1849.
55. Ibid.
56. Nigel Leask, '"Wandering through Elbis": Absorption and Containment in Romantic Exoticism', in *Romanticism and Colonialism: Writing and Empire, 1789–1830*, ed. Tim Fulford and Peter J. Kitson (Cambridge: Cambridge University Press, 1998), 165–88 at p. 173.
57. *Athenaeum*, 16 February 1850.
58. 'Minor Topics of the Month', *Art Journal*, March 1850.
59. 'The Drama and Public Amusements', *Critic*, 1 May 1850.
60. 'The Drama and Public Amusements', *Critic*, 15 November 1850.
61. 'Panorama of the Arctic Regions', *Era*, 24 February 1850.

62. 'Burford's Panorama of the Arctic Regions', *Morning Chronicle*, 11 February 1850.
63. 'Panorama of the Arctic Regions', *Observer*, 10 February 1850.
64. James McGuire and James Quinn, eds., *Dictionary of Irish Biography* (Cambridge: Cambridge University Press, 2009), http://dib.cambridge.org.
65. *Oxford Dictionary of National Biography*, www.oxforddnb.com.
66. Many letters, dating from 1818 to 1836, exist between Lieutenant James Browne and the Commissioners of Howth Harbour, beginning with: James Browne, Application for the position of harbour master at Howth, 1 June 1818, OPW/8/HOW/2686, Howth Harbour 1800–52, National Archives of Ireland, http://www.nationalarchives.ie. Boase tells us that William Henry Browne was the 'son of Mr. Browne, harbour master at Dublin'. Frederic Boase, *Modern English Biography* (Truro: Netherton & Worth, 1892–1921), 1:447.
67. A Constant Traveller, 'To the Proprietor of the Freeman's Journal', *Freeman's Journal*, 30 April 1825.
68. Boase, *Modern English Biography*, 1:447; William O'Byrne, *A Naval Biographical Dictionary: Comprising the Services of All Living Naval Officers, Admirals, Captains, Commanders, Lieutenants, Masters, Engineers, Mates, Second-masters, Chaplains, Naval Instructors, Medical officers, Paymasters, and Assistant-Paymasters, together with those of the Officers of the Royal Marines*, 3 vols. (London: John Murray, 1861), 1:134.
69. For example, William Henry Browne, *Baseelan Island and Part of Mindanao Island* [1844], ADM 344/1459, Admiralty Records, TNA UK.
70. William Henry Browne, *Ten Coloured Views Taken during the Arctic Expedition* (London: Ackermann, 1850).
71. Officers and Seamen of the Expedition, *Arctic Miscellanies: A Souvenir of the Late Polar Search*, ed. James Donnet (London: Colburn & Co., 1852), 66.
72. SPRI, Cambridge and NMM, Greenwich.
73. Burford, 'Summer and Winter Views', 3.
74. Burford, 'Summer and Winter Views', 3.
75. 'Some Account of an Extraordinary Traveller', *Household Words*, 20 April 1850.
76. Burford, 'Summer and Winter Views', 3.
77. Ibid.
78. Edmund Burke, *A Philosophical Enquiry into the Origin of Our Ideas of the Sublime and the Beautiful*, ed. Adam Phillips (Oxford: Oxford University Press, 1990), 77.
79. *Arctic Miscellanies*, 195.
80. Edward Belcher, *The Last of the Arctic Voyages*, 2 vols. (London: Lovell Reeve, 1855), 1:213. Here, Belcher refers to the 'cheerless' reputation of the long Arctic winter that was apparently much commented upon in Britain.

81. 'Burford's Arctic Panorama', *Illustrated London News*, 16 February 1850.
82. 'Minor Topics of the Month', *Art Journal*, March 1850.
83. 'Panorama of the Polar Regions', *Literary Gazette*, 18 February 1850.
84. 'Panorama of the Arctic Regions', *Observer*, 10 February 1850.
85. 'Notes of the Month', *Gentleman's Magazine*, March 1850.
86. William Henry Browne, *Valley of the Glaciers* [1848], PAH0066, NMM, Greenwich. While the composition of the panorama is the same as this sketch, Burford may well have used additional drawings by Browne and his prior knowledge of Arctic and Alpine subject matter to enhance the scene.
87. J. D. Gilpin, 'Outline of the Voyage of H.M.S. Enterprize and Investigator to Barrow Strait in Search of Sir John Franklin', *Nautical Magazine* 19 (January 1850): 8–19; 82–90; 160–70; 230 at p. 15. Gilpin records that later a gale was blowing, causing the ships to become endangered. Another picture by Edward Adams, *Glacier Bay, copied from Lieut Brown by candle light*, showing a similar scene, exists at SPRI. www.spri.cam.ac.uk/museum/catalogue/article/y83.11.12/.
88. A nunatak is 'an isolated peak of rock projecting above the surface of inland ice' (*Oxford English Dictionary*). The three peaks represented in Browne's painting are now 'marginal rock islands at the icecap periphery'. Mauri Pelto, 'Upernavik Glacier Retreat, NW Greenland', *From a Glaciers* [sic] *Perspective*, https://glacierchange.wordpress.com/2011/09/17/upernavik-glacier-retreat-nw-greenland.
89. William Henry Browne, *Ten Coloured Views Taken during the Arctic Expedition* (London: Ackermann, 1850). The set of lithographs can be viewed on the Royal Collection Trust website: www.rct.uk/collection/search#/1/collection/750930/arctic-expedition-in-search-of-sir-john-franklin-1848-49.
90. How much control Browne would have had over the production of the lithographs is unclear, but, as a lieutenant, he may have had little say in their rendering, and the Admiralty may have controlled their publication.
91. 'A Visit to the Arctic Discovery Ships', *Household Words*, 20 April 1850.
92. 'Arctic Exploration', *New Monthly Magazine*, February 1860.
93. Burford, 'Summer and Winter Views', 3.
94. Simon Schama, *Landscape and Memory* (London: HarperCollins, 1995), 451–2.
95. Burford, 'Summer and Winter Views', 4.
96. Jen Hill, *White Horizon: The Arctic in the Nineteenth-Century British Imagination* (Albany: State University of New York Press, 2008), 76.
97. Andrew Lambert, *Franklin: Tragic Hero of Polar Navigation* (London: Faber & Faber, 2009), 196; 219; 296.
98. Garrison, 'Virtual Reality', 14–15.

99. Kirsten Hvenegård-Lassen, 'Journey to the Centre of the Ice: Narrating Ice-Core Drillings in Northern Greenland', *Studies in Travel Writing* 20, no. 3 (September 2016): 237–48 at p. 241.
100. Stefan Jonsson, 'Where Humanism Finds its Ends: Lessons from Pia Arke and Katarina Pirak Sikku on the Difficulty of Narrating the Arctic', *Studies in Travel Writing* 20, no. 3 (September 2016): 226–36 at p. 229.
101. Although I did not locate an obvious prototype sketch for the winter view by Browne, it is possible that it could be in private ownership.
102. 'Sir James Ross' Arctic Expedition', *Illustrated London News*, 17 November 1849.
103. Ibid.
104. Although there are no Inuit shown in the representation, it is important to note that the expedition, unlike those approaching from the Pacific or overland from the south, wintered in an area far from Indigenous settlements.
105. *Illustrated London News*, 23 February 1850.
106. Heather Davis-Fisch, *Loss and Cultural Remains in Performance: The Ghosts of the Franklin Expedition* (New York: Palgrave Macmillan, 2012), 44.
107. John Matthews, The Journal of John Matthews from September 1850 to April 1855, written during a voyage in search of the Franklin Expedition, [1850–5] mg N07/11D: Entry for May 1851. Monograph Collection, RGS, London.
108. [William Chimmo], *Euryalus; Tales of the Sea, a Few Leaves from the Diary of a Midshipman* (London: J.D. Potter, 1860), 300–1.
109. Sherard Osborn, *Stray Leaves from an Arctic Journal* (London: Longman, Brown, Green, and Longmans, 1852), 168–9.
110. Osborn, *Stray Leaves*, 169.
111. William Henry Browne, *Coast of N. Somerset* [1849], N: 956, SPRI, Cambridge.
112. S. Renou, *Delineations Physical, Intellectual, and Moral, Exemplifying the Philosophy of Christianity* (London: Shaw and Sons, 1838), 193–4.
113. 'Panorama of the Arctic Regions, Leicester Square', *Era*, 24 February 1850.
114. Burford, 'Summer and Winter Views', 9.
115. Catherine Lanone, 'Arctic Spectacles in *Jane Eyre* and *Villette*', *Brontë Studies* 34, no. 2 (July 2009): 117–26 at pp. 120; 118.
116. [Chimmo], *Euryalus*, 306; 308–9.
117. Ibid., 309
118. William Simpson, Diary Written on board HMS Plover [January 1848–December 1850], JOD/76: Entry for 28 August 1849, NMM, Greenwich.
119. *Illustrated London News*, 9 February 1850.

120. 'Some Account of an Extraordinary Traveller', *Household Words*, 20 April 1850.
121. Goliah Muff [William Thackeray], 'The Sights of London', *Punch*, 6 April 1850.
122. 'Panorama of the Polar Regions', *The Literary Gazette: A Weekly Journal of Literature, Science, and the Fine Arts*, 16 February 1850.
123. 'Panorama of the Polar Regions', *Observer*, 10 February 1850.
124. 'Panorama of the Arctic Regions', *Era*, 24 February 1850.
125. Burford, 'Summer and Winter Views', 4. Coruscations referred to the glimmering light of the aurora.
126. Burford, 'Summer and Winter Views', 14.
127. 'Burford's Panorama of the Polar Regions', *Illustrated London News*, 23 February 1850.
128. 'Burford's Panorama of the Polar Regions', *Illustrated London News*, 23 February 1850.
129. Shane McCorristine, '"Involuntarily We Listen": Hearing the Aurora Borealis in Nineteenth-Century Arctic Exploration and Science', *Canadian Journal of History* 48 (Spring–Summer 2013): 29–61 at p. 29.
130. Neil Bone, *Aurora: Observing and Recording Nature's Spectacular Light Show* (New York: Springer, 2007), 139.
131. Denison Olmsted, 'On the Recent Secular Period of the Aurora Borealis', *Smithsonian Contributions to Knowledge* 8, no. 3 (Washington DC: Smithsonian Institution, 1856), 1–52 at p. 52.
132. J. E. P., 'Aurora Borealis', *National Magazine* (October 1830), 461–3.
133. Lambert, *Franklin: Tragic Hero*, 166.
134. J. F. W. Herschel, 'Meteorology', in *A Manual of Scientific Enquiry: Prepared for the Use of Her Majesty's Navy and Adapted for Travellers in General. Published by Authority of the Lords Commissioners of the Admiralty*, ed. John F. W. Herschel (London: John Murray, 1849), 316.
135. Logbook HMS Enterprize [1848–9], ADM/55/44, Admiralty Records, TNA UK.
136. Ibid., Entry for 28 November 1848.
137. Ibid., Entry for 24 January 1849.
138. Gilpin, 'Outline of the Voyage', 86–7.
139. James Clark Ross, 'Narrative of the Proceedings of Captain Sir James C. Ross, in Command of the Expedition through Lancaster Sound and Barrow Straits', *Arctic Blue Books*, 1850a.
140. Osborn, *Stray Leaves*, 164.
141. Belcher, *Arctic Voyages*, 1:174. In 1850, due to the more southerly location of the magnetic north pole, the aurora would have been more visible at lower latitudes than it is today.

142. Ibid., 180. The aurora is still often referred to as female in twentieth-century popular discourse.
143. Walter W. May, Journal of HMS Assistance [20 April 1852–19 August 1853], Walter May Fonds MG24-H31: Entry for 10 December 1852, LAC, Ottawa.
144. May, Journal of HMS Assistance, Entry for 11 December 1852.
145. William Parry noted, on his third Arctic voyage to discover the Northwest Passage (1824–5), that it was only on the return journey to England in autumn, as the ship sailed to more southerly latitudes in Davis Strait and across the Atlantic, that the aurora was frequent and spectacular: 'The next brilliant display . . . which far surpassed anything of the kind observed at Port Bowen [in the Arctic], occurred on the night of the 24th of September, in latitude 58$^1/_2$.' This line of latitude crosses the north of mainland Scotland. Parry described the display at length and referred to the aurora's 'undulating motion . . . called the "merry dancers", which are seen in beautiful perfection at the Shetland Islands'. *Journal of a Third Voyage for the Discovery of a North-West Passage* (Philadelphia: H. C. Carey and I. Lea, 1826), 148–9.
146. McCorristine, '"Involuntarily We Listen"', 31.
147. 'Panorama of the Arctic Regions', *Era*, 24 February 1850.
148. 'Panorama of the Arctic Regions', *Observer*, 10 February 1850.
149. William Jerden (ed.), 'Panorama of the Polar Regions', *The Literary Gazette: A Weekly Journal of Literature, Science, and the Fine Arts*, 16 February 1850.

5 'Power and Truth'

1. Samuel Gurney Cresswell, *War, Ice and Piracy: The Remarkable Career of a Victorian Sailor: The Journals and Letters of Samuel Gurney Cresswell*, ed. Dominick Harrod (London: Chatham Publishing, 2000), 107.
2. Samuel Gurney Cresswell, *A Series of Eight Sketches in Colour . . . of the Voyage of H.M.S. 'Investigator'* (London: Day and Ackermann, 1854).
3. Advert, *Publishers' Circular*, 1 August 1854; Cresswell, *War, Ice and Piracy*, 118.
4. 'Art', *Critic*, 1 March 1850.
5. Book Review, *Art Journal*, April 1850.
6. The cost of the lithographs ranged from sixteen shillings to two pounds and two shillings per set and is discussed further on p. 146.
7. William Henry Browne, *Ten Coloured Views Taken during the Arctic Expedition* (London: Ackermann, 1850).
8. Cresswell, A Series of Eight Sketches in Colour, 1854; Walter William May, *A Series of Fourteen Sketches Made during the Voyage up Wellington Channel* (London: Day and Son, 1855).
9. 'The Art of Lithography', *Morning Chronicle*, 18 January 1855.

10. 'The North-West Passage', *Morning Chronicle*, 24 October 1853.
11. Cian Duffy, *The Landscapes of the Sublime, 1700–1830: Classic Ground* (Houndmills, Basingstoke: Palgrave Macmillan, 2013), 21.
12. B. E. Maidment, *Reading Popular Prints, 1790–1870*, 2nd ed. (Manchester: Manchester University Press, 2001), 11.
13. Bernard Smith, *Imagining the Pacific: In the Wake of the Cook Voyages* (New Haven: Yale University Press, 1992), 54.
14. Michael Clarke, *The Concise Oxford Dictionary of Art Terms*, 2nd ed., with Deborah Clarke (Oxford: Oxford University Press, 2010), 145.
15. Michael Twyman, 'The Illustration Revolution', in *The Cambridge History of the Book in Britain*, ed. David McKitterick, vol. 6, *1830–1914* (Cambridge: Cambridge University Press, 2009), 117–43 at p. 121.
16. Richard C. Sha, *The Visual and Verbal Sketch in British Romanticism* (Philadelphia: University of Pennsylvania Press, 1998), 1; 4; 12.
17. Tim Youngs, ed., *Travel Writing in the Nineteenth Century: Filling in the Blank Spaces* (London: Anthem Press, 2006), 9.
18. Martin J. S. Rudwick, 'The Emergence of a Visual Language for Geological Science 1760–1840', *History of Science* 14 (1976): 149–95 at p. 156.
19. Naomi Hume, 'The Nature Print and Photography in the 1850s', *Photography* 35, no. 1 (2011): 44–58 at p. 54; Gunilla Törnvall, 'From Copperplate to Color Lithograph: On the Modernisation of an Illustrated Flora 1800–1900', *Book History* 20 (2017): 126–49 at p. 137.
20. Erika Piola, 'The Rise of Early American Lithography and Antebellum Visual Culture', *Winterthur Portfolio* 48, no. 2/3 (2014): 125–38 at p. 127.
21. Michael Twyman, *Lithography 1800–1850: The Techniques of Drawing on Stone in England and France and Their Application in Works of Topography* (Oxford: Oxford University Press, 1970), 161.
22. Advertisement, *Athenaeum*, 23 February 1850.
23. *Publishers' Circular*, 1 August 1854. The eight views, with a map, were printed in 'coloured tints'.
24. 'The North-West Passage', *Morning Chronicle*, 24 October 1853.
25. 'List of New Books', *Literary Gazette*, 23 June 1855.
26. Arthur L. Bowley, *Wages in the United Kingdom in the Nineteenth Century* (Cambridge: Cambridge University Press, 1900), 78; 73.
27. 'The Art of Lithography', *Morning Chronicle*, 18 January 1855. The lithographs do not appear to have been reproduced as engravings in periodicals.
28. Advertisement, *Morning Chronicle*, 24 October; Advertisement, *Athenaeum*, 14 January 1854.
29. All the lithographs from this folio are available to view in high resolution on the Royal Collection Trust website: www.rct.uk/collection/search#/1/collection/750930/arctic-expedition-in-search-of-sir-john-franklin-1848–49.

30. Browne, *Ten Coloured Views*, 5.
31. The author of the text is anonymous.
32. Advertisement, *Athenaeum*, 23 February 1850.
33. Advertisement, *Athenaeum*, 23 February 1850. The advertisement in the *Critic* on 1 March used the same text.
34. 'Art', *Critic*, 1 March 1850.
35. 'Ten Coloured Views Taken during the Arctic Expedition', *Athenaeum*, 2 March 1850.
36. 'Art', *Critic*, 1 March 1850.
37. Ibid.
38. Ibid.
39. Book Review, *Art Journal*, April 1850.
40. 'Ten Coloured Views Taken during the Arctic Expedition', *Athenaeum*, 2 March 1850.
41. 'Ten Coloured Views', *Athenaeum*, 2 March 1850.
42. William Henry Browne, letter to the editor, *The Times*, 29 December 1849; *Literary Gazette*, 5 January 1850; *Athenaeum*, 12 January 1850. (See also Chapter 4, footnote 6.)
43. The fact that one lithograph caption, in error, refers to Leopold Island instead of Somerset Island also indicates that Browne was not supervising the work of the printers.
44. Lithography was associated with the production of knowledge, particularly in sciences such as natural history. Moreover, their expensive price raised them above popular culture.
45. The connections between the watercolour *Valley of the Glaciers, Greenland* and the lithograph *Great Glacier, Near Uppernavik* are discussed in Chapter 4.
46. P. R. Dawes and R. L. Christie, 'Geomorphic Regions', in *Geology of the Innuitian Orogen and Arctic Platform of Canada and Greenland*, ed. H. P. Trettin (Ottawa: Geological Survey of Canada, 1991), 27–56 at pp. 40–4.
47. Rebecca Bedell, *The Anatomy of Nature: Geology and American Landscape Painting, 1825–75* (Princeton: Princeton University Press, 2001), 4; 15.
48. Henry Briant, 'A Catalogue of the Library Established on board HMS Assistance' [1853], Arctic Pamphlets vol. 3, mg N07/08K, RGS, London.
49. The house, renovated and expanded, still exists today and trades as King Sitric Restaurant and Accommodation.
50. Aaron Clarke, Sarah Gatley, and Matthew Parkes, 'Fingal – County Geological Site Report', *The Geological Heritage of Fingal* (Dublin: Geological Survey of Ireland, 2007).
51. British Parliament, *Arctic Blue Books*, 1852b, 68; 71; 74–5.

52. Review of *Ten Coloured Views*, by William Henry Browne, *Art Journal*, April 1850.
53. Edmund Burke, *A Philosophical Enquiry into the Origin of Our Ideas of the Sublime and the Beautiful*, ed. Adam Phillips (Oxford: Oxford University Press, 1990), 79.
54. William Henry Browne, *Coast of N. Somerset* [1849], N: 956, SPRI, Cambridge.
55. Deniz Tekiner, *Modern Art and the Romantic Vision* (Lanham: University Press of America, 2000), 11; 15.
56. Kate Flint, *The Victorians and the Visual Imagination* (Cambridge: Cambridge University Press, 2000), 120.
57. This lithograph is referred to as 'Prince Regent's Inlet' in the list of titles at the start of the bound version of the set of lithographs, but the title 'The Sledges Arriving at the Southern Depôt' is used on the lithograph itself.
58. 'Art', *Critic*, 1 March 1850.
59. Ann C. Colley, *Victorians in the Mountains: Sinking the Sublime* (London: Routledge, 2016), 57.
60. Ibid., 68.
61. Review of *Ten Coloured Views*, by William Henry Browne, *Art Journal*, April 1850.
62. L. H. Neatby, 'Robert J.L. McClure (1807–1873)', *Arctic* 38, no. 1 (March 1985): 70–1 at p. 70. The McClure expedition was the first to cross the passage, but they did not officially 'navigate' it; this was not achieved until the Norwegian Roald Amundsen's Gjøa expedition (1903–6).
63. L. H. Neatby, 'Samuel Gurney Cresswell (1827–1867)', *Arctic* 35, no. 4 (Dec 1982): 554–5 at p. 554.
64. I am very grateful to Chloe Phillips at Norfolk Record Office for photographing Cresswell's work, as I was unable to travel to the archives personally. The pictures are contained in an album.
65. Review of *Series of Eight Sketches*, by Samuel Gurney Cresswell, *Art Journal*, September 1854.
66. Ibid.
67. 'Fine Arts', *Morning Chronicle*, 25 September 1855.
68. 'Fine Arts', *Athenaeum*, 16 September 1854.
69. 'Fine Arts', *Athenaeum*, 16 September 1854.
70. Cresswell, *War, Ice and Piracy*, 57.
71. The set of lithographs in its entirety is available to view online courtesy of TPL: https://static.torontopubliclibrary.ca/da/pdfs/408043.pdf.
72. Samuel Gurney Cresswell, *Brown's Island, Coast of America* [1850], WMH 3-1-D4-1, NRO, Norwich.

73. Samuel Gurney Cresswell, *Discovery of Barings Island September 6th 1850* [1850], WMH 3–1-D4-1, NRO, Norwich.
74. Not the expedition member of the same name.
75. Review of *Series of Eight Sketches*, by Samuel Gurney Cresswell, *Art Journal*, September 1854.
76. Ibid.
77. 'Fine Arts', *Athenaeum*, 16 September 1854.
78. Review of *Series of Eight Sketches*, by Samuel Gurney Cresswell, *Art Journal*, September 1854.
79. Samuel Gurney Cresswell, *Position of H.M.S. Investigator after Heavy Pressure, 1852* [1852], WMH 3–1-D4-1, NRO, Norwich.
80. Advertisement, *Morning Chronicle*, 24 October 1853. Inglefield's narrative *A Summer Search* had been published in June 1853, and in December the Gallery of Illustration was showing five 'Polar views . . . the whole, or nearly so, being taken from the spirited sketches of Captain Inglefield'. The views were accompanied by a lecture; the whole exhibition was 'new and delightful. Everyone seemed charmed by it.' 'Entertainments for Christmas', *Examiner*, 24 December 1853.
81. 'The North-West Passage', *Athenaeum*, 29 October 1853.
82. Inglefield's lithographs can be viewed on the Royal Museums Greenwich website: https://collections.rmg.co.uk/collections/objects/156627.html.
83. See Introduction, footnote 74. I have come across this print on seven book covers to date.
84. Alex Armstrong, *A Personal Narrative of the Discovery of the North-West Passage; with Numerous Incidents of Travel and Adventure during nearly Five Years' Continuous Service in the Arctic Regions while in Search of the Expedition under Sir John Franklin* (London: Hurst and Blackett, 1857), 420.
85. Review of *Series of Eight Sketches*, by Samuel Gurney Cresswell, *Art Journal*, September 1854.
86. McClure had ordered all journals and many other possessions to be left aboard when the ship was abandoned in 1853. Johann Miertsching, *Frozen Ships: The Arctic Diary of Johann Miertsching 1850–1854*, trans. L. H. Neatby (Toronto: Macmillan, 1967), 192. Thus, recollections like Miertsching's were written after the expedition, and no journal by Cresswell exists for the period while he was on the *Investigator* from 1850 until the abandonment of the ship.
87. Miertsching, *Frozen Ships*, 134.
88. Cresswell, *War, Ice and Piracy*, 76.
89. Armstrong, *Personal Narrative*, 405–6.
90. 'H.M.S. "Investigator", in the Arctic Regions', *Illustrated London News*, 29 October 1853.

91. 'H.M.S. "Investigator", in the Arctic Regions', *Illustrated London News*, 29 October 1853.
92. Ibid.
93. Samuel Gurney Cresswell, *Position of H.M.S. Investigator, Sept 19, 1851* [1851], WMH 3–1-D4-1, NRO, Norwich.
94. Armstrong, *Personal Narrative*, 442–3.
95. Delia Millar, 'William Simpson (1823–1899)', in *Oxford Dictionary of National Biography* (Oxford: Oxford University Press, 2004), www.oxforddnb.com/view/article/25597.
96. Clements R. Markham, *The Arctic Navy List 1773–1873* (1875; repr., Dallington: Naval and Military Press, 1992), 35.
97. Sherard Osborn, 'Queen's Illuminated Magazine and North Cornwall Gazette'. Published in Winter Quarters, Arctic Regions [28 October 1852–12 February 1853]: Add MS 35305, Barrow Bequest, vol. 6, Western Manuscripts, BL.
98. 'Fine Arts', *Morning Chronicle*, 11 September 1850.
99. May, *Fourteen Sketches*, 3.
100. See, for example, Sherard Osborn, *Stray Leaves from an Arctic Journal* (London: Longman, Brown, Green, and Longmans, 1852), 145–6.
101. Walter W. May, Journal of HMS Assistance [20 April 1852–19 August 1853], Walter May Fonds MG24-H31: Entry for 2 December 1852, LAC, Ottawa.
102. 'Fine Arts', *Morning Chronicle*, 11 September 1855.
103. May, *Fourteen Sketches*, 4.
104. 'List of New Books', *Literary Gazette*, 23 June 1855.
105. May, *Fourteen Sketches*, 7.
106. PAF7055; PAF7056; PAF7060, NMM, Greenwich.
107. May, *Fourteen Sketches*, 4.
108. Walter William May, *Assistance in Tow of the Pioneer* [1853], PAF7061, NMM, Greenwich.
109. Walter William May, *Sledge Party Returning through Water in the Month of July* [1853/4], PAF7058, NMM, Greenwich. The painting is available to view on the Royal Museums Greenwich website: https://collections.rmg.co.uk/collections/objects/101885.html.
110. May, *Fourteen Sketches*, 6.
111. Robert G. David, *The Arctic in the British Imagination 1818–1914* (Manchester: Manchester University Press, 2000), 44–5.
112. Walter William May, *Loss of the McLellan* [1853/4], PAF7059, NMM, Greenwich; Walter William May, *II.—Loss of the McLellan* (London: Day & Son, 1855). The lithograph can be viewed on the Royal Museums Greenwich website: https://collections.rmg.co.uk/collections/objects/110604.html.

113. 'Destruction of Fourteen Vessels by Ice-Bergs in the Atlantic–Upwards of One Hundred Lives Lost', *Kerry Examiner*, 21 June 1850, Irish Newspaper Archive.
114. May, *Fourteen Sketches*, 3.
115. George Frederick McDougall, *The Eventful Voyage of H. M. Discovery Ship 'Resolute' to the Arctic Regions* (London: Longman, Brown, Green, Longmans, & Roberts, 1857), 64–5.
116. Osborn, 'Queen's Illuminated Magazine', 3 [January 1853]: 1.
117. Osborn, 'Queen's Illuminated Magazine', 3 [January 1853]: 1. 'Piling the agony' is a phrase that Osborn also used in *Stray Leaves from an Arctic Journal* (1852) and may have an American origin. It indicates an excessive use of ornamentation and adjectives in written work, what may be termed purple prose today: 'A tendency towards what the Americans term "piling the agony" is the besetting sin of poets these days.' *Northern Tribune*, 1854, 169.
118. May, Journal, Entry for 15 July 1852.
119. Walter William May, *Division of Sledges Passing Cape Lady Franklin* [1853/4], PAF7055, NMM, Greenwich; Walter William May, *X.—Division of Sledges Passing Cape Lady Franklin* (London: Day & Son, 1855). The latter is available to view on the Royal Museums Greenwich website: https://collections.rmg.co.uk/collections/objects/137982.html.
120. 'Fine Arts', *Morning Chronicle*, September 11, 1850.
121. John Barrell, *The Dark Side of the Landscape: The Rural Poor in English Painting 1730–1840* (Cambridge: Cambridge University Press, 1989), 21.
122. An Admiralty publication on Inuit vocabulary included the phrase 'I want to buy twelve good dogs / Well trained to draw sledges'. John Washington, *Eskimaux and English Vocabulary, for the Use of the Arctic Expeditions, Published by Order of the Lords Commissioners of the Admiralty* (London: John Murray, 1850), 104.
123. William Barr, 'The Use of Dog Sledges during the British Search for the Missing Franklin Expedition in the North American Arctic Islands, 1848–59', *Arctic* 62, no. 3 (September 2009): 257–72 at p. 267.
124. Osborn, 'Queen's Illuminated Magazine', 1 [November 1852]: 24.
125. Ibid., 4 [February 1853]: 25.
126. Barr, 'Dog Sledges', 271.
127. Andrew Lambert, *Franklin: Tragic Hero of Polar Navigation* (London: Faber & Faber, 2009), 322; 332.
128. Lisa Bloom, *Gender on Ice: American Ideologies of Polar Expeditions* (Minneapolis: University of Minnesota Press, 1993), 120.
129. Duffy, *Landscapes of the Sublime*, 105.

130. Carl Thompson, *The Suffering Traveller and the Romantic Imagination* (Oxford: Clarendon Press, 2007), 3.
131. Ibid., 25.
132. Duffy, *Landscapes of the Sublime*, 117.
133. Hume, 'The Nature Print and Photography', 54.

6 Conclusion

1. Lisa Bloom, *Gender on Ice: American Ideologies of Polar Expeditions* (Minneapolis: University of Minnesota Press, 1993), 6.
2. Walter W. May, Journal of HMS Assistance [20 April 1852–19 August 1853], Walter May Fonds MG24-H31: Entry for 11 October 1852, LAC, Ottawa.
3. Edward Belcher, *The Last of the Arctic Voyages* (London: Lovell Reeve, 1855), 2: 7.
4. Sherard Osborn, *Stray Leaves from an Arctic Journal* (London: Longman, Brown, Green, and Longmans, 1852), 280.
5. Isaac Israel Hayes, *An Arctic Boat Journey, in the Autumn of 1854*, 2nd ed. (Boston: Ticknor and Fields, 1867), 331.
6. Carolyn Strange, 'Reconsidering the "Tragic" Scott Expedition: Cheerful Masculine Home-making in Antarctica, 1910–1913', *Journal of Social History* 48, no. 1 (Fall 2012): 66–88 at p. 67.
7. Richard C. Powell, 'Inscribing the North: The Historical Geographies of Circumpolar Description' (paper, International Conference of Historical Geographers, Royal Geographical Society, London, 5 July 2015).
8. Keith Battarbee and John Erik Fossum, 'NACS and the Arctic' (lecture, Nordic Association of Canadian Studies XI, University of Turku, Finland, 14 August 2015).
9. Robert Burford, 'Description of Summer and Winter Views of the Polar Regions as Seen during the Expedition of Capt. James Clark Ross' (London, 1850), 4.
10. Michael Byers, *Who Owns the Arctic? Understanding Sovereignty Disputes in the North* (Vancouver: Douglas and McIntyre, 2009), 48–9.
11. Ibid., 50.
12. Adriana Craciun, *Writing Arctic Disaster: Authorship and Exploration* (Cambridge: Cambridge University Press, 2016), 225.
13. Ibid., 230.
14. [Unknown Artist], *North Star Mount in Wolstenholme Sound, SE. 1.8 miles: land of a brownish purplish tinge, August 15th 1850* [1850], T3262 / nla.obj-133007737, NLA, Canberra. A colour version can be viewed in high resolution on the National Library of Australia's website: https://nla.gov.au/nla.obj-133007737/view.
15. [William Chimmo], *Euryalus; Tales of the Sea, a Few Leaves from the Diary of a Midshipman* (London: J.D. Potter, 1860), 218.

Bibliography

Primary Sources

Periodicals Cited

Unless otherwise indicated, all periodicals were published in London. Where consulted online, the collection is listed following the periodical title.

Art Journal – British Periodicals, Proquest

Athenaeum – British Periodicals, Proquest *Chambers's Journal of Popular Literature, Science and Arts* (Edinburgh)

Critic – British Periodicals, Proquest

Era – British Library Newspapers, Gale

Examiner – British Periodicals, Proquest

Freeman's Journal (Dublin) – Irish News Archive

Gentleman's Magazine – British Periodicals, Proquest

Household Words

Illustrated London News – Illustrated London News Archive, Gale

Kerry Examiner (Tralee) – Irish News Archive

Literary Gazette – British Periodicals, Proquest

Lloyd's Weekly Newspaper – British Library Newspapers, Gale

London Journal

Morning Chronicle – British Library Newspapers, Gale

Musical World – British Periodicals, Proquest

National Magazine – British Periodicals, Proquest

Nautical Magazine

New Monthly Magazine – British Periodicals, Proquest

New Sporting Magazine – British Periodicals, Proquest

Northern Tribune (Newcastle-on-Tyne) – Google Books

Observer – Proquest Historical Newspapers

Publishers' Circular – Literature Online, Proquest

Punch, or the London Charivari	– Punch Historical Archive, Gale
The Times	– The Times Digital Archive
Theatrical Journal	– British Periodicals, Proquest

Manuscript and Picture Sources

Adams, Edward. Collection. Scott Polar Research Institute, University of Cambridge.

Admiralty Hydrography Records. The National Archives, London.

Arctic Pamphlets. Vol. 3. Royal Geographical Society, London.

Barrow, John. Bequest. British Library.

Cator, John Bertie. Collection. Scott Polar Research Institute, University of Cambridge.

Illustration Archive. Cardiff University. https://illustrationarchive.cf.ac.uk.

Kivell, Rex Nan. Collection. National Library of Australia, Canberra. http://catalogue.nla.gov.au.

McCormick, Robert. Collection. Wellcome Library, London. (Viewed on microfilm at Library and Archives Canada, Ottawa.)

McDougall, George Frederick. Collection. Library and Archives Canada, Ottawa.

Manuscripts Collection. National Library of Ireland, Dublin.

Markham, Clements Robert. Collection. Royal Geographical Society, London.

Matthews, John. Collection. Royal Geographical Society, London.

May, Walter William. Fonds. Library and Archives Canada, Ottawa.

Mumford, William T. Fonds. Library and Archives Canada, Ottawa.

Ommanney, Erasmus. Collection. Royal Geographical Society, London.

Polar Art Collection. Scott Polar Research Institute, University of Cambridge. http://www.spri.cam.ac.uk/museum/catalogue/polarart/gallery/.

Prints and Drawings. National Maritime Museum, Greenwich.

Snow, William Parker. Collection. Royal Geographical Society, London.

Toronto Public Library. https://www.torontopubliclibrary.ca/.

Webster, John Clarence. Canadiana Collection. New Brunswick Museum, Saint John. www.nbm-mnb.ca.

Printed Sources

Sources originally written or published in the eighteenth or nineteenth centuries but reprinted later are included here.

Arctic Blue Books Online: British Parliamentary Papers on Exploration in the Canadian North. 1818–78. Indexed by Andrew Taylor. Winnipeg: University of Manitoba, 2003. https://www.umanitoba.ca/libraries/units/archives/collections/subject/arcticstudies/arcticbb/.

The English Catalogue of Books from 1835 to 1863. London: Sampson Low, 1864.

A Select Collection of Songs. Newcastle Upon Tyne, 1806.

The Universal Songster; or, Museum of Mirth. Vol. 2. London, 1826.

Adams, Edward. *Into the Savage Land: The Alaskan Journal of Edward Adams*. Edited by Ernest Sipes. Surrey, BC: Hancock House, 2007.

'Notes on the Birds of Michalaski, Norton Sound'. *Ibis, a Quarterly Journal of Ornithology* 2, no. 8 (October 1878): 420–2.

Agassiz, Louis. *Études sur les Glaciers*. Neuchatel: Jent et Gassmann, 1840.

Armstrong, Alex. *A Personal Narrative of the Discovery of the North-West Passage; with Numerous Incidents of Travel and Adventure during nearly Five Years' Continuous Service in the Arctic Regions while in Search of the Expedition under Sir John Franklin*. London: Hurst and Blackett, 1857.

Back, George. *Narrative of an Expedition in H.M.S. Terror, Undertaken with a View to Geographical Discovery on the Arctic Shores, in the Years 1836–7*. London: John Murray, 1838.

Baillie, John. *An Essay on the Sublime* (1747). In *The Sublime: A Reader in British Eighteenth-Century Aesthetic Theory*, ed. Andrew Ashfield and Peter de Bolla, 87–100. Cambridge: Cambridge University Press, 1996.

Barker, Henry. 'Description of a View of the North Coast of Spitzbergen. Now Exhibiting in the Large Rotunda of Henry Aston Barker's Panorama, Leicester Square'. London, 1820.

Bayly, Thomas Haynes. *The Poetical Works of Thomas Haynes Bayly*, ed. Helena Beecher Haynes, 2 vols. London: Richard Bentley, 1844.

Beattie, James. *The Minstrel; or, the Progress of Genius*. Bk. 1. London: Dilly, 1771.

Belcher, Edward. *The Last of the Arctic Voyages; Being a Narrative of the Expedition in H.M.S.* Assistance, *under the Command of Sir Edward* Belcher, *C.B., in Search of Sir John* Franklin, *during the Years 1852–53–54*. 2 vols. London: Lovell Reeve, 1855.

Narrative of the Voyage of H. M. S. Samarang: during the Years 1843–46: Employed Surveying the Islands of the Eastern Archipelago. 2 vols. London: Reeve, Benham, and Reeve, 1848.

Bellot, Joseph René. *Memoirs of Lieutenant Joseph René Bellot, . . . with his Journal of a Voyage in the Polar Seas, in Search of Sir John Franklin*. 2 vols. London: Hurst and Blackett, 1855.

Boase, Frederic. *Modern English Biography*. 6 vols. Truro: Netherton and Worth, 1892–1921.

Bowley, Arthur L. *Wages in the United Kingdom in the Nineteenth Century*. Cambridge: Cambridge University Press, 1900.

Bray, Emile de. *A Frenchman in Search of Franklin: De Bray's Arctic Journal 1852–54*, trans. and ed. William Barr. Toronto: University of Toronto Press, 1992.

British Parliament. 'Papers and Correspondence relative to the Arctic Expedition under Sir John Franklin'. In Taylor, *Arctic Blue Books*, 1848a, 7. https://www.umanitoba.ca/libraries/units/archives/collections/subject/arcticstudies/arcticbb/.

Browne, William Henry. *Ten Coloured Views Taken during the Arctic Expedition of Her Majesty's Ships 'Enterprise' and 'Investigator' under the Command of Sir James Clark Ross*. London: Ackermann, 1850.

Burford, Robert. 'Description of a View of the Continent of Boothia, Discovered by Captain Ross, in His Late Expedition to the Polar Regions, Now Exhibiting at the Panorama, Leicester Square'. London, 1834.

'Description of Summer and Winter Views of the Polar Regions as Seen during the Expedition of Capt. James Clark Ross'. London, 1850.

Burke, Edmund. *A Philosophical Enquiry into the Origin of Our Ideas of the Sublime and the Beautiful*, 1757, ed. with an introduction and notes by Adam Phillips. Oxford: Oxford University Press, 1990.

[Chimmo, William]. *Euryalus; Tales of the Sea, a Few Leaves from the Diary of a Midshipman*. London: J.D. Potter, 1860.

Collinson, Richard. *Journal of H.M.S. Enterprise on the Expedition in Search of Sir John Franklin's Ships by Behring Strait. 1850–55*, ed. T. B. Collinson. London: Sampson Low, Marston, Searle, and Rivington, 1889.

Combe, William and Thomas Rowlandson. *The Tour of Doctor Syntax in Search of the Picturesque*. London: Ackermann, 1812.

Cook, James. *A Voyage towards the South Pole and round the World*. London: W. Strahan and T. Cadell, 1777.

Cresswell, Samuel Gurney. *A Series of Eight Sketches in Colour . . . of the Voyage of H.M.S. 'Investigator' (Captain McClure) during the Discovery of the North West Passage*. London: Day and Ackermann, 1854.

War, Ice and Piracy: The Remarkable Career of a Victorian Sailor: The Journals and Letters of Samuel Gurney Cresswell, ed. Dominick Harrod. London: Chatham Publishing, 2000.

Cruikshank, George. *George Cruikshank's Table-Book*, ed. Gilbert Abbott. London: Punch, 1845.

Dana, R. H. *Two Years before the Mast: A Personal Narrative of Life at Sea*. 4th ed. London: Edward Moxon, 1845.

Diebitsch-Peary, Josephine. *My Arctic Journal: A Year among Ice-Fields and Eskimos*, 1893. Reprint, New York: Copper Square Press, 2002.

Don, William Gerard. 'Reminiscences of the Baltic Fleet of 1855'. Brechin, 1894.

Field, George. *Rudiments of the Painters' Art; or a Grammar of Colouring, Applicable to Operative Painting, Decorative Architecture, and the Arts*. London: John Weale, 1850.

Gilpin, J. D. 'Outline of the Voyage of H.M.S. Enterprize and Investigator to Barrow Strait in search of Sir John Franklin', *Nautical Magazine* 19 (1850): 8–19; 82–90; 160–70; 230.

Gilpin, William. *Observations, on Several Parts of England, Particularly the Mountains and Lakes of Cumberland and Westmoreland*. London: Cadell and Davies, 1808.

Three Essays: On Picturesque Beauty; On Picturesque Travel; and On Sketching Landscape: to which is Added a Poem on Landscape Painting. London: Blamire, 1794.

Godfrey, William C. *Godfrey's Narrative of the Last Grinnell Arctic Exploring Expedition, in Search of Sir John Franklin, 1853–4–5*. Philadelphia: J. T. Lloyd, 1857.

Goodsir, Robert. *An Arctic Voyage to Baffin's Bay and Lancaster Sound, in Search of Friends with Sir John Franklin*. London: John van Voorst, 1850.

Hayes, Isaac Israel. *An Arctic Boat Journey in the Autumn of 1854*. London: Richard Bentley, 1860.

An Arctic Boat Journey, in the Autumn of 1854. 2nd ed. Boston: Ticknor and Fields, 1867.

Hearne, Samuel. *A Journey from Prince of Wales's Fort in Hudson Bay, to the Northern Ocean*. London: A. Strahan and T. Cadell, 1795.

Herschel, John F. W., ed. *A Manual of Scientific Enquiry: Prepared for the Use of Her Majesty's Navy and Adapted for Travellers in General. Published by Authority of the Lords Commissioners of the Admiralty*. London: John Murray, 1849.

Hogarth, William. *The Analysis of Beauty. Written with a View of Fixing the Fluctuating Ideas of Taste*, 1753. New ed., London: W. Strahan, 1772.

Hooper, William Hulme. *Ten Months among the Tents of the Tuski, with Incidents of an Arctic Boat Expedition in Search of Sir John Franklin, as far as the Mackenzie River, and Cape Bathurst*. London: John Murray, 1853.

Humboldt, Alexander von. *Cosmos: Sketch of a Physical Description of the Universe*. Vol. 2, translated under the superintendence of Edward Sabine. London: Longman, Brown, Green, and Longmans and John Murray, 1849.

Hunt, Robert. *Photography: A Treatise on the Chemical Changes Produced by Solar Radiation, and the Production of Pictures from Nature, by the Daguerreotype, Calotype, and Other Photographic Processes*. London: Griffin, 1851.

Inglefield, E. A. *A Summer Search for Sir John Franklin with a Peep into the Polar Basin*. London: Thomas Harrison, 1853.

Kane, Elisha Kent. *Arctic Explorations: The Second Grinnell Expedition in Search of Sir John Franklin, 1853, '54, '55*. 2 vols. Philadelphia: Childs and Petersen, 1856.

The U.S. Grinnell Expedition in Search of Sir John Franklin. A Personal Narrative. New York: Harper & Brothers, 1854.

Kennedy, William. *A Short Narrative of the Second Voyage of the Prince Albert in Search of Sir John Franklin*. London: W. H. Dalton, 1853.

Lane, Robert. *Practical Reflections for Every Day throughout the Year*. 3rd ed., ed. Edward Peach. London: Keating and Brown, 1835.

Maguire, Rochfort. *The Journal of Rochfort Maguire, 1852–1854: Two Years at Point Barrow, Alaska, aboard HMS Plover in the Search for Sir John Franklin*, ed. John R. Bockstoce. 2 vols. London: Hakluyt Society, 1988.

Markham, Clements R. *Franklin's Footsteps: A Sketch of Greenland, along the Shores of which His Expedition Passed, and of the Parry Isles, where the Last Traces of It were Found*. London: Chapman and Hall, 1853.

Life of Sir Admiral Leopold McClintock. London: John Murray, 1909.

The Arctic Navy List 1773 – 1873, 1875. Reprint, Dallington: Naval and Military Press, 1992.

May, Walter William. *A Series of Fourteen Sketches Made during the Voyage up Wellington Channel in Search of Sir John Franklin and the Missing Crews of H.M. Discovery-Ships Erebus and Terror*. London: Day and Son, 1855.

McClintock, Francis. *The Voyage of the 'Fox' in the Arctic Seas: A Narrative of the Discovery of the Fate of Sir John Franklin and His Companions*. London: John Murray, 1859.

The Voyage of the 'Fox' in the Arctic Seas in Search of Franklin and His Companions. 3rd ed. London: John Murray, 1869.

McClure, Robert. *The Discovery of the North-West Passage*, ed. Sherard Osborn. London: Longman, Brown, Green, Longmans, & Roberts, 1856.

McCormick, Robert. *Narrative of a Boat Expedition up the Wellington Channel in the Year 1852, under the Command of R. McCormick, R.N., F.R.C.S., in H.M.B. 'Forlorn Hope', in Search of Sir John Franklin*. London: Eyre and Spottiswoode, 1854.

McDougall, George Frederick. *The Eventful Voyage of H. M. Discovery Ship 'Resolute' to the Arctic Regions*. London: Longman, Brown, Green, Longmans, & Roberts, 1857.

Miertsching, Johann. *Frozen Ships: The Arctic Diary of Johann Miertsching 1850–1854*, trans. L. H. Neatby. Toronto: Macmillan, 1967.

Mitchell, Robert. *Plans and Views in Perspective of Buildings Erected in England and Scotland*. London: Wilson and Co., 1801.

Noble, R. P. *A Guide to Water Colour Painting, with an Illustration in Colour*. London: George Rowney, 1850.

O'Byrne, William R. *A Naval Biographical Dictionary: Comprising the Services of All Living Naval Officers, Admirals, Captains, Commanders, Lieutenants, Masters, Engineers, Mates, Second-masters, Chaplains, Naval Instructors, Medical officers, Paymasters, and Assistant-Paymasters, together with those of the Officers of the Royal Marines*. New ed., 3 vols. London: John Murray, 1861.

Officers and Seamen of the Expedition. *Arctic Miscellanies: A Souvenir of the Late Polar Search*, ed. James Donnet. London: Colburn & Co., 1852.

Olmsted, Denison. 'On the Recent Secular Period of the Aurora Borealis'. *Smithsonian Contributions to Knowledge* 8, no. 3 (Washington, DC: Smithsonian Institution, 1856): 1–52.

Ord, John Walker. *The Bard, and Minor Poems*, ed. John Lodge. 2nd ed. London: Simpkin, Marshall & Co., 1842.

Osborn, Sherard. *Stray Leaves from an Arctic Journal; or, Eighteen Months in the Polar Regions, in Search of Sir John Franklin's Expedition, in the Years 1850–51*. London: Longman, Brown, Green, and Longmans, 1852.

Osborn, Sherard and George F. McDougall. *Illustrated Arctic News*. London: Ackermann, 1852.

Parry, William. *Journal of a Third Voyage for the Discovery of a North-West Passage*. Philadelphia: H. C. Carey and I. Lea, 1826.

Phipps, Constantine. *A Voyage towards the North Pole*. London: Bowyer and Nichols, 1774.

Price, Uvedale. *An Essay on the Picturesque, as Compared with the Sublime and the Beautiful; and, on the Use of Studying Pictures, for the Purpose of Improving Real Landscape*. London: J. Robson, 1796.

Radcliffe, Ann. *A Journey Made in the Summer of 1794, through Holland and the Western Frontier of Germany*. 2 vols. London: G. G. and J. Robinson, 1795.

The Mysteries of Udolpho: A Romance; Interspersed with Some Pieces of Poetry. 4 vols. 4th ed. London: G. G. and J. Robinson, 1799.

'The Snow-Fiend'. In *Gaston de Blondeville and St. Alban's Abbey*, vol. 4. 4 vols. London: Henry Colburn, 1826.

Renou, S. *Delineations Physical, Intellectual, and Moral, Exemplifying the Philosophy of Christianity*. London: Shaw and Sons, 1838.

Richardson, John. *Arctic Searching Expedition: A Journal of a Boat Voyage through Ruperts Land and the Arctic Sea in Search of Sir John Franklin*. 2 vols. London: Longman, Brown, Green, and Longmans, 1851.

Ross, James Clark. 'Narrative of the Proceedings of Captain Sir James C. Ross, in Command of the Expedition through Lancaster Sound and Barrow Straits'. In Taylor, *Arctic Blue Books*, 1850a. http://www.umanitoba.ca/libraries/units/archives/collections/subject/arcticstudies/arcticbb/viewbb.php?t=1850a.

Ross, John. *A Voyage of Discovery: Made under the Orders of the Admiralty, in His Majesty's Ships Isabella and Alexander, for the Purpose of Exploring Baffin's Bay, and Inquiring into the Probability of a North-West Passage*. London: John Murray, 1819.

Ruskin, John. 'Præterita. Outlines of Scenes and Thoughts Perhaps Worthy of Memory in my Past Life', 1886–9. In *Library Edition of the Works of John Ruskin*, 39 vols., vol. 35, ed. E. T. Cook and Alexander Wedderburn. London: George Allen, 1908.

'Deucalion, and Other Studies in Rocks and Stones', 1878. In *Library Edition of the Works of John Ruskin*, 39 vols., vol. 26, ed. E. T. Cook and Alexander Wedderburn. London: George Allen, 1906.

Seeman, Berthold. *Narrative of the Voyage of H.M.S. Herald during the Years 1845–51, under the Command of Captain Henry Kellett, R.N., C.B.; being a Circumnavigation of the Globe, and Three Cruizes to the Arctic Regions in Search of Sir John Franklin*. 2 vols. London: Reeve, 1853.

'Flora of Western Eskimaux-Land'. In *The Botany of the Voyage of H.M.S. Herald under the Command of Captain Henry Kellett, R.N., C.B., during the Years 1845–51*, 11–56. London: Reeve, 1852.

Shelley, Mary Wollstonecraft. *Frankenstein: or, The Modern Prometheus – the 1818 Text*. Reprint, ed. Marilyn Butler. Oxford: Oxford University Press, 2008.

Smyth, W. H. *An Alphabetical Digest of Nautical Terms*, ed. Edward Belcher. London: Blackie and Son, 1867.

Snow, William Parker. *Voyage of the Prince Albert in Search of Sir John Franklin: A Narrative of Every-Day Life in the Arctic Seas*. London: Longman, Brown, Green, and Longmans, 1851.

Southey, Robert. *The Life of Nelson*. London: Collins, 1813.

Stevenson, H. 'The Late Mr. Edward Adams'. *Ibis, a Quarterly Journal of Ornithology* 2, no. 8 (October 1878): 420–42.

Sutherland, Peter C. *Journal of a Voyage in Baffin's Bay and Barrow Straits . . . in Search of the Missing Crews of H.M. Ships Erebus and Terror*. 2 vols. London: Longman, Brown, Green, and Longmans, 1852.

Taylor, Tom. *Leicester Square; Its Associations and its Worthies*. London: Bickers and Son, 1875.

Thornthwaite, William. *A Guide to Photography, Containing Simple and Practical Details . . . Including Photogenic Drawing, Calotype, Daguerreotype*. London: Horn, Thornthwaite, and Wood, 1845.

Verne, Jules. *The Adventures of Captain Hatteras*, trans. unknown. London: Ward Lock, 1876.

Washington, John. *Eskimaux and English Vocabulary, for the Use of the Arctic Expeditions. Published by Order of the Lords Commissioners of the Admiralty*. London: John Murray, 1850.

Secondary Sources

Agnew, John A. *Place and Politics: The Geographical Mediation of State and Society*. Boston and London: Allen and Unwin, 1987.

Place and Politics in Modern Italy. Chicago: University of Chicago Press, 2002.

Alasuutari, Pertti. *Qualitative Method and Cultural Studies*. London: Sage, 1995.

Altick, Richard D. *Punch: The Lively Youth of a British Institution*. Columbus: Ohio State University Press, 1997.

The Shows of London: A Panoramic History of Exhibitions, 1600–1862. Cambridge, MA: Belknap Press of Harvard University Press, 1978.

Anderson, Patricia. 'Illustration'. In *Victorian Periodicals and Victorian Society*, ed. J. Don Vann and Rosemary T. VanArsdel, 127–42. Aldershot: Scolar, 1994.

Andrews, Malcolm. *The Picturesque: Literary Sources and Documents*, ed. Graham Clarke. Vol. 1, *The Idea of the Picturesque and the Vogue for Scenic Tourism*. Mountfield: Helm Information, 1994.

Armstrong, Isobel. *Victorian Glassworlds: Glass Culture and the Imagination 1830–1880*. Oxford: Oxford University Press, 2008.

Arnold, Dana. *Art History: A Very Short Introduction*. Oxford: Oxford University Press, 2004.

Atwood, Margaret. *Strange Things: The Malevolent North in Canadian Literature*. Oxford: Clarendon Press, 1995.

Barasch, Moshe. *Modern Theories of Art: From Winckelmann to Baudelaire*. London: New York University Press, 1990.

Barr, William. *Arctic Hell-Ship: The Voyage of HMS Enterprise 1850–55*. Edmonton: University of Alberta Press, 2007.

'The Use of Dog Sledges during the British Search for the Missing Franklin Expedition in the North American Arctic Islands, 1848–59'. *Arctic* 62, no. 3 (September 2009): 257–72.

Barrell, John. *The Dark Side of the Landscape: The Rural Poor in English Painting 1730–1840*. Cambridge: Cambridge University Press, 1989.

Barringer, Tim. 'Images of Otherness and the Visual Production of Difference: Race and Labour in Illustrated Texts, 1850–1865'. In *The Victorians and Race*, ed. Shearer West, 34–52. Aldershot: Scolar Press, 1996.

Barry, Peter. *Beginning Theory: An Introduction to Literary and Cultural Theory*. Manchester: Manchester University Press, 2009.

Barthes, Roland. *Mythologies*. London: Vintage, 1993.

Battarbee, Keith and John Erik Fossum. 'NACS and the Arctic'. Lecture given at the Nordic Association of Canadian Studies XI, University of Turku, Finland, August 2015.

Beattie, Owen and John Geiger. *Frozen in Time: The Fate of the Franklin Expedition* London: Bloomsbury, 1987.

Bedell, Rebecca. *The Anatomy of Nature: Geology and American Landscape Painting, 1825–75*. Princeton: Princeton University Press, 2001.

Behrisch, Erika. '"Far as the Eye Can Reach": Explorers' Poetry in the Arctic 1832–1852'. In 'Science and Victorian Poetry', ed. Gowan Dawson and Sally Shuttleworth, special issue, *Victorian Poetry* 41, no. 1 (Spring 2003): 73–92.

'Voices of Silence, Texts of Truth: Imperial Discourse and Cultural Negotiations in Nineteenth-Century British Arctic Exploration Narrative'. PhD thesis, Queen's University, Kingston, 2002.

Behrisch Elce, Erika. '"Adapted for Travellers in General": En Route with the British Admiralty's Manual of Scientific Enquiry.' *Journeys* 15, no. 1 (2014): 1–22.

'"One of the Bright Objects That Solace Us in These Regions": Labour, Leisure, and the Arctic Shipboard Periodical, 1820–1852'. *Victorian Periodicals Review* 46, no. 3 (Fall 2013): 343–67.

Behrisch Elce, Erika, ed. *As Affecting the Fate of My Absent Husband: Selected Letters of Lady Franklin concerning the Search for the Lost Franklin Expedition, 1848–1860*. Montreal and Kingston: McGill-Queen's University Press, 2009.

Bell, Clive. 'The Aesthetic Hypothesis'. In *Art in Theory 1900–1990*, ed. Charles Harrison and Paul Wood, 113–16. Oxford: Blackwell, 1992.

Bell, Leonard. 'Seeing Differently: Shifts from the Pictorial to the Photographic in the Shaping of the South Pacific'. *Journal for Maritime Research* 7, no. 1 (March 2005): 16–25.

Belyea, Barbara. 'Captain Franklin in Search of the Picturesque'. *Essays on Canadian Writing* 40 (Spring 1990): 1–24.

Bermingham, Ann. *Learning to Draw: Studies in the Cultural History of a Polite and Useful Art*. New Haven and London: Yale University Press, 2000.

Bershad, Sonia S. 'The Drawings and Watercolours by Rear Admiral Frederick William Beechey, F.R.S., P.R.G.S. (1796–1856) in the Collection of the Arctic Institute of North America, University of Calgary'. *Arctic* 33, no. 1 (March 1980): 117–67.

Birkwood, Susan. 'From "Naked Country" to "Sheltering Ice": Rudy Wiebe's Revisionist Treatment of John Franklin's First Arctic Narrative'. *Nordlit* 23 (2008): 25–38.

Bleichmar, Daniela. 'The Imperial Visual Archive: Images, Evidence, and Knowledge in the Early Modern Hispanic World'. *Colonial Latin America Review* 24, no. 2 (2015): 236–66.

Bloom, Lisa. *Gender on Ice: American Ideologies of Polar Expeditions*. Minneapolis: University of Minnesota Press, 1993.

Bloom, Michelle E. *Waxworks: A Cultural Obsession*. Minneapolis: University of Minnesota Press, 2003.

Blum, Hester. 'The News at the Ends of the Earth: Polar Periodicals'. In *Unsettled States: Nineteenth-Century American Literary Studies*, ed. Dana Luciano and Ivy G. Wilson, 158–88. New York: New York University Press, 2014.

The News at the Ends of the Earth: The Print Culture of Polar Exploration. Durham, NC: Duke University Press, 2019.

Bolla, Peter de. 'Toward the Materiality of Aesthetic Experience'. *Diacritics* 32, no. 1 (Spring 2002): 19–37.

Bone, Neil. *Aurora: Observing and Recording Nature's Spectacular Light Show*. New York: Springer, 2007.

Boucher, Ellen. 'Arctic Mysteries and Imperial Ambitions: The Hunt for Sir John Franklin and the Victorian Culture of Survival'. *The Journal of Modern History* 90 (March 2018): 40–75.

Bourdieu, Pierre. *Distinction: A Social Critique of the Judgement of Taste*. Translated by Richard Nice. Cambridge, MA: Harvard University Press, 1984.

Bouwers, Eveline G. *Public Pantheons in Revolutionary Europe: Comparing Cultures of Remembrance, c. 1790–1840*. New York: Palgrave Macmillan, 2012.

Boxwell, David A. 'The Follies of War: Cross-Dressing and Popular Theatre on the British Front Lines, 1914–18'. *Modernism/Modernity* 9, no. 1 (January 2002): 1–20.

Brienen, Rebecca Parker. *Visions of a Savage Paradise: Albert Eckhout, Court Painter in Colonial Dutch Brazil*. Amsterdam: Amsterdam University Press, 2006.

Briggs, Asa. *Victorian Things*. Stroud: Sutton Publishing, 2003.

Bryson, Norman, Michael Ann Holly, and Keith Moxley, eds. *Visual Theory: Painting and Interpretation*. Cambridge: Polity in association with Blackwell, 1991.

Bullough, Vern L. and Bonnie L. Bullough. *Cross Dressing, Sex, and Gender*. Philadelphia: University of Pennsylvania Press, 1993.

Bumsted, John Michael. *The Peoples of Canada: A Pre-Confederation History*. Toronto: Oxford University Press, 1992.

Burke, Peter. *What is Cultural History?* Cambridge: Polity, 2004.

Byers, Michael. *International Law and the Arctic*. Cambridge: Cambridge University Press, 2013.

Who Owns the Arctic? Understanding Sovereignty Disputes in the North. Vancouver: Douglas and McIntyre, 2009.

Canadian Broadcasting Corporation. 'Arctic Rescue Fears Loom as Massive Cruise Ship Prepares to Sail Northwest Passage'. Includes an interview with Michael Byers. *CBC News*, April 2016. www.cbc.ca/news/canada/north/cruise-ships-safety-northwest-passage-1.3518712.

Canadian Coast Guard. 'Ice Climatology and Environmental Conditions'. In *Ice Navigation in Canadian Waters*, 33–78 (2012). https://www.ccg-gcc.gc.ca/publications/icebreaking-deglacage/ice-navigation-glaces/docs/ice-navigation-dans-les-galces-eng.pdf.

Carey, Daniel. 'Anthropology's Inheritance: Renaissance Travel, Romanticism and the Discourse of Identity'. *History and Anthropology* 14, no. 2 (2003): 107–26.

'Compiling Nature's History: Travellers and Travel Narratives in the Early Royal Society'. *Annals of Science* 54, no. 3 (1997): 269–92.

'Truth, Lies, and Travel Writing'. In *The Routledge Companion to Travel Writing*, ed. Carl Thompson, 3–14. London: Routledge, 2016.

Casey, Edward S. *Getting Back into Place: Toward a Renewed Understanding of the Place-World.* Bloomington and Indianapolis: Indiana University Press, 1993.

Cavell, Janice. 'Comparing Mythologies: Twentieth-Century Constructions of Sir John Franklin'. In *Canadas of the Mind: The Making and Unmaking of Canadian Nationalisms in the Twentieth Century*, ed. Norman Hillmer and Adam Chapnick, 15–45. Montreal and Kingston: McGill-Queen's University Press, 2007.

'Going Native in the North: Reconsidering British Attitudes during the Franklin Search, 1848–59'. *Polar Record* 45 (2009): 25–35.

Letter to the Editor. *Arctic* 63, no. 2 (June 2010): 249–51.

'Making Books for Mr Murray: The Case of Edward Parry's Third Arctic Narrative'. *The Library* 14, no. 1 (March 2013): 45–69.

'Publishing Sir John Franklin's Fate: Cannibalism, Journalism, and the 1881 Edition of Leopold McClintock's *Voyage of the "Fox" in the Arctic Seas*'. *Book History* 16 (2013): 155–84.

Review of *Writing Arctic Disaster: Authorship and Exploration*, by Adriana Craciun. *Journal of Historical Geography*. Published electronically, 25 September 2016: 1–2.

'The Polar Regions'. In Thompson, *Routledge Companion to Travel Writing*, 425–34.

Tracing the Connected Narrative: Arctic Exploration in British Print Culture, 1818–1860. Toronto: University of Toronto Press, 2008.

Chilvers, Ian. *The Concise Oxford Dictionary of Art and Artists.* Oxford: Oxford University Press, 2003.

Clarke, Aaron, Sarah Gatley, and Matthew Parkes. 'Fingal – County Geological Site Report'. In *The Geological Heritage of Fingal.* Dublin: Geological Survey of Ireland, 2007. https://www.gsi.ie/NR/rdonlyres/31939C0C-8E81-4109-8C49-04BA1368BA13/0/Fingal_section2_part2.pdf.

Clarke, Michael. *The Concise Oxford Dictionary of Art Terms.* 2nd ed., with Deborah Clarke. Oxford: Oxford University Press, 2010.

Close, Susan. 'The Camera and the Contact Zone: Re-Envisioning the Representation of Aboriginal Women in the Canadian North'. In Codell, *Transculturation in British Art, 1770–1930*, 111–24.

Codell, Julie F. 'The Art of Transculturation'. In Codell, *Transculturation in British Art, 1770–1930*, 1–17.

Codell, Julie F., ed. *Transculturation in British Art, 1770–1930*. London: Routledge, 2016.

Colley, Ann C. *Victorians in the Mountains: Sinking the Sublime*. London: Routledge, 2016.

Connelly, Charlotte and Claire Warrior. 'Survey Stories in the History of British Polar Exploration: Museums, Objects, People'. *Notes and Records* 73 (2019): 259–74.

Cordingly, David. *Marine Painting in England 1700–1900*. London: Studio Vista, 1974.

Cosgrove, Denis. *Social Formation and Symbolic Landscape*. Madison: University of Wisconsin Press, 1998.

Costello, Timothy M., ed. *The Sublime: From Antiquity to the Present*. Cambridge: Cambridge University Press, 2012.

Craciun, Adriana. 'The Franklin Relics in the Arctic Archive'. *Victorian Literature and Culture* 42 (2014): 1–31.

'What is an Explorer?' *Eighteenth-Century Studies* 45, no. 1 (Fall 2011): 29–51.

Writing Arctic Disaster: Authorship and Exploration. Cambridge: Cambridge University Press, 2016.

'Writing the Disaster: Franklin and *Frankenstein*'. *Nineteenth-Century Literature* 65, no. 4 (March 2011): 433–80.

Crary, Jonathan. *Techniques of the Observer: On Vision and Modernity in the Nineteenth Century*. Cambridge, MA: MIT Press, 1992.

Cresswell, Tim. *Place: A Short Introduction*. Oxford: Blackwell, 2004.

Crookshank, Anne and Desmond Fitzgerald, the Knight of Glin. *The Watercolours of Ireland: Works on Paper in Pencil, Pastel and Paint c. 1600–1914*. London: Barrie and Jenkins, 1994.

Cyriax, Richard. *Sir John Franklin's Last Arctic Expedition: The Franklin Expedition. A Chapter in the History of the Royal Navy*. London: Methuen, 1939.

D'Alleva, Anne. *Methods and Theories of Art History*. London: Laurence King Publishing, 2005.

Davey, James. 'Singing for the Nation: Balladry, Naval Recruitment and the Language of Patriotism in Eighteenth-Century Britain'. *Mariner's Mirror* 103, no. 1 (2017): 43–66.

David, Robert G. *The Arctic in the British Imagination 1818–1914*. Manchester: Manchester University Press, 2000.

Davidson, Peter. *The Idea of North*. London: Reaktion, 2005.

Davis, Richard C. 'Fact and Fancy in History and Biography: The Case of Greenstockings'. *Polar Record* 37 (January 2001): 5–12.

'History or His/Story? The Explorer Cum Author'. *Studies in Canadian Literature* 16, no. 2 (1991): 93–111.

'Vision and Revision: John Franklin's Arctic Landscapes'. *Australian-Canadian Studies* 6, no. 2 (1989): 23–33.

Davis-Fisch, Heather. *Loss and Cultural Remains in Performance: The Ghosts of the Franklin Expedition*. New York: Palgrave Macmillan, 2012.

Dawes, P. R. and R. L. Christie. 'Geomorphic Regions'. In *Geology of the Innuitian Orogen and Arctic Platform of Canada and Greenland*, ed. H. P. Trettin, 27–56. Ottawa: Geological Survey of Canada, 1991.

Day, Alan. 'Furthest North'. *Library Review* 47, no. 1 (1998): 42–4.

Delgado, James P. *Across the Top of the World: The Quest for the Northwest Passage*. London: British Museum Press, 1999.

Demuth, Bathsheba. *Floating Coast: An Environmental History of the Bering Strait*. New York: W. W. Norton, 2020.

Despoix, Philippe. 'The Exchanged Portrait and the Lethal Picture: Visualization Techniques and Native Knowledge in Samuel Hearne's Sketches from His Trek to the Arctic Ocean and John Webber's Record of the Northern Pacific'. *Eighteenth-Century Fiction* 23, no. 4 (Summer 2011): 667–89.

Dickason, Olive Patricia. *Canada's First Nations: A History of Founding Peoples from the Earliest Times*. Oxford: Oxford University Press, 2002.

Doel, Ronald E., Urban Wråkberg, and Suzanne Zeller. 'Science, Environment, and the New Arctic'. *Journal of Historical Geography* 44 (2014): 2–14.

Donald, Diana. 'The Arctic Fantasies of Edwin Landseer and Briton Riviere: Polar Bears, Wilderness and Notions of the Sublime'. *Tate Papers* 13 (April 2010). www.tate.org.uk/research/publications/tate-papers/arctic-fantasies-edwin-landseer-and-briton-riviere-polar-bears.

Douglas, Mary. 'The Social Control of Cognition: Some Factors in Joke Perception'. *Man* 3, no. 3 (September 1968): 361–76.

Driver, Felix. 'Distance and Disturbance: Travel, Exploration and Knowledge in the Nineteenth Century'. *Transactions of the Royal Historical Society* 14 (December 2004): 73–92.

Geography Militant: Cultures of Exploration and Empire. Oxford: Blackwell, 2001.

'Material Memories of Travel: The Albums of a Victorian Naval Surgeon'. *Journal of Historical Geography* 69 (July 2020): 32–54.

Driver, Felix and Luciana Martins, eds. *Tropical Visions in an Age of Empire*. Chicago: University of Chicago Press, 2005.

Driver, Felix and Luciana Martins. 'Views and Visions of the Tropical World'. In Driver and Martins, *Tropical Visions*, 3–20.

Duffy, Cian. *The Landscapes of the Sublime, 1700–1830: Classic Ground*. Houndmills, Basingstoke: Palgrave Macmillan, 2013.

Dunbar, Moira J. 'The Royal Arctic Theatre, with Drawings from The Illustrated Arctic News Published on Board H.M.S. Resolute in the Canadian Arctic, 31 October 1850 to 14 March 1851'. *Canadian Art* 15, no. 2 (1958): 110–13.

Durey, Michael. 'Exploration at the Edge: Reassessing the Fate of Sir John Franklin's Last Arctic Expedition'. *The Great Circle* 30, no. 2 (2008): 3–40.

Dutton, Jacqueline. 'Imperial Ice? The Influence of Empire on Contemporary French and British Antarctic Travel Writing'. *Studies in Travel Writing* 13, no. 4 (December 2009): 369–80.

Eber, Dorothy Harley. *Encounters on the Passage: Inuit Meet the Explorers*. Toronto: University of Toronto Press, 2008.

Eco, Umberto. *Faith in Fakes: Travels in Hyperreality*. London: Vintage, 1998.

Edwards, Elizabeth. 'Photography and the Performance of History'. *Kronos* no. 27 (November 2001): 15–29.

Raw Histories: Photographs, Anthropology and Museums. Oxford: Berg, 2001.

Elden, Stuart. *Understanding Henri Lefebvre: Theory and the Possible*. London: Continuum, 2004.

Emmerson, Charles. *The Future History of the Arctic: How Climate, Resources and Geopolitics are Reshaping the North, and Why It Matters to the World*. London: Vintage Books, 2011.

Energy, Mines and Resources Canada. *The National Atlas of Canada 5th Edition*. Ottawa: Energy, Mines and Resources Canada, 1985. http://geoscan.nrcan.gc.ca/starweb/geoscan/servlet.starweb?path=geoscan/fulle.web&search1=R=293986.

Falkenhausen, Susanne von. *Beyond the Mirror: Seeing in Art History and Visual Culture Studies*, trans. Nicholas Grindell. Bielefeld: Transcript, 2020.

Ferguson, Frances. *Solitude and the Sublime*. New York and London: Routledge, 1992.

Fleming, Fergus. *Barrow's Boys*. London: Granta Books, 1998.

Flint, Kate. *The Victorians and the Visual Imagination*. Cambridge: Cambridge University Press, 2000.

Freedgood, Elaine. *Victorian Writing about Risk*. Cambridge: Cambridge University Press, 2000.

Friel, Ian. *Maritime History of Britain and Ireland c. 400–2001*. London: British Museum Press, 2003.

Fry, Roger. *Vision and Design*. Harmondsworth: Penguin, 1940.

Fulford, Tim. 'Romanticism and Colonialism: Races, Places, Peoples, 1800–30'. In Fulford and Kitson, *Romanticism and Colonialism*, 35–47.

Fulford, Tim and Peter J. Kitson, eds. *Romanticism and Colonialism: Writing and Empire, 1780–1830*. Cambridge: Cambridge University Press, 1998.

Garrison, Laurie. 'Virtual Reality and Subjective Responses: Narrating the Search for the Franklin Expedition through Robert Burford's Panorama'. *Early Popular Visual Culture* 10, no. 1 (February 2012): 7–22.

Garrison, Laurie, ed. *Panoramas, 1787–1900: Texts and Contexts*, 5 vols. London: Pickering and Chatto, 2013.

Gascoigne, Bamber. *How to Identify Prints*. London: Thames and Hudson, 1998.

Gilroy, Amanda. 'Introduction'. In Gilroy, *Romantic Geographies*, 1–15.

Gilroy, Amanda, ed. *Romantic Geographies: Discourses of Travel 1775–1844*. Manchester: Manchester University Press, 2000.

Gombrich, Ernst H. *Art and Illusion: a Study in the Psychology of Pictorial Representation*. Princeton: Princeton University Press, 2000.

Government of Canada. 'Arctic Charting'. In *Fisheries and Oceans Canada*. Last modified 11 June 2021. http://www.charts.gc.ca/arctic-arctique/index-eng.asp.

Granger, Frank, ed. *A Treatise on the Sublime by Longinus*. London: Stanley Nott, 1935.

Green, Anna. *Cultural History*. Basingstoke: Palgrave Macmillan, 2008.

Greppi, Claudio. '"On the Spot": Traveling Artists and the Iconographic Inventory of the World, 1769–1859'. In Driver and Martins, *Tropical Visions*, 23–42.

Hansson, Heidi. 'Introduction: The Arctic in Literature for Children and Young Adults'. In Hansson, Lindgren Leavenworth, and Ryall, *The Arctic in Literature*, 5–26.

'Staging the Arctic 1819–1909 and 2014'. *Nordlit* 35 (2015): 47–62.

'The Gentleman's North: Lord Dufferin and the Beginnings of Arctic Tourism'. *Studies in Travel Writing* 13, no. 1 (February 2009): 61–73.

Hansson, Heidi, Maria Lindgren Leavenworth, and Anka Ryall, eds. *The Arctic in Literature for Children and Young Adults*. London: Routledge, 2020.

Harding, Richard. *Modern Naval History: Debates and Prospects*. London: Bloomsbury Academic, 2016.

Harley, J. B. 'Historical Geography and the Cartographic Illusion'. *Journal of Historical Geography* 15, no. 1 (1989): 80–91.

Harper, J. Russell. *Painting in Canada: A History*. Toronto: University of Toronto Press, 1978.

Harrington, Austin. *Art and Social Theory: Sociological Arguments in Aesthetics*. Cambridge: Polity, 2004.

Heath, Malcolm. 'Longinus and the Ancient Sublime'. In Costello, *The Sublime*, 11–23.

Heus, Ronald, Hein A. M. Daanen, and George Havenith. 'Physiological Criteria for Functioning of Hands in the Cold'. *Applied Ergonomics* 26, no. 1 (1995): 5–13.

Hill, Jen. *White Horizon: The Arctic in the Nineteenth-Century British Imagination*. Albany: State University of New York Press, 2008.

Hoag, Elaine. 'Caxtons of the North: Mid-Nineteenth Century Arctic Shipboard Printing'. *Book History* 4 (2001): 81–114.

'Shipboard Printing on the Franklin Search Expeditions: A Bibliographical Study of the *Plover* Press'. *Papers of the Bibliographical Society of Canada* 38, no. 1 (2000): 7–71.

Houston, C. Stuart. 'Robert Hood (1797–1821)'. *Arctic* 36, no. 2 (June 1983): 210–11.

Houston, C. Stuart and Ian S. MacLaren. *Arctic Artist: The Journal and Paintings of George Back, Midshipman with Franklin, 1819–1822*. Montreal and Kingston: McGill-Queen's University Press, 1994.

Høvik, Ingeborg. 'Arctic Images 1818–1859'. PhD thesis, University of Edinburgh, 2013.

'Framing the Arctic: Reconsidering Roald Amundsen's Gjøa Expedition Imagery'. *Nordlit* 25 (2015): 137–60.

'Girlhood in the Arctic: Word-Image Relations in R. M. Ballantyne's Canadian Adventures', in Hansson, Lindgren Leavenworth, and Ryall, *The Arctic in Literature*, 89–106.

'Heroism and Imperialism in the Arctic: Edwin Landseer's Man Proposes – God Disposes'. *Nordlit* no. 23 (2008): 183–94.

'Reproducing the Indigenous: John Møller's Studio Portraits of Greenlanders in Context'. *Acta Borealia* 33, no. 2 (2016): 166–88.

Howard, Peter. *Landscapes: The Artists' Vision*. London: Routledge, 1991.

Howells, Richard and Joaquim Negreiros. *Visual Culture*. Cambridge: Polity, 2012.

Hubbard, Phil et al. *Key Thinkers on Space and Place*. London: Sage, 2004.

Huggan, Graham and Lars Jensen, eds. *Postcolonial Perspectives on the European High Arctic: Unscrambling the Arctic*. London: Palgrave Macmillan, 2016.

Hume, Naomi. 'The Nature Print and Photography in the 1850s'. *Photography* 35, no. 1 (2011): 44–58.

Huneault, Kristina. 'Miniature Objects of Cultural Covenant: Portraits and First Nations Sitters in British North America'. *Canadian Art Review* 30, no. 1/2 (2005): 87–100.

Huntington, Henry and Gunter Weller. 'Introduction'. In *Arctic Climate Impact Assessment*, ed. Carolyn Symon, 1–20. Cambridge: Cambridge University Press, 2005.

Hvenegård-Lassen, Kirsten. 'Journey to the Centre of the Ice: Narrating Ice-Core Drillings in Northern Greenland'. *Studies in Travel Writing* 20, no. 3 (September 2016): 237–48.

Hyde, Ralph. 'Burford, Robert (1791–1861)'. *Oxford Dictionary of National Biography*. Oxford: Oxford University Press, 2004. http://www.oxforddnb.com/view/article/3965?docPos=2.

Panoramania! The Art And Entertainment of the 'All-Embracing' View. London: Trefoil in association with the Barbican Art Gallery, 1988.

Inglis, David and John Hughson, eds. *The Sociology of Art: Ways of Seeing*. Basingstoke: Palgrave Macmillan, 2005.

Inuit Circumpolar Council. *Inuit Circumpolar Council Charter*. Nuuk, 2010. http://inuit.org/en/icc-greenland/inuit-circumpolar-council-charter/.

Inuit Tapiriit Kanatami, 'Inuit Regions of Canada' (Ottawa, 2019). https://www.itk.ca/about-canadian-inuit/.

Ivins, William M. *Prints and Visual Communication*. London: Routledge and Kegan Paul, 1953.

Jackson, Christine E. 'M & N Hanhart: Printers of Victorian Music Covers'. *Fontes Artis Musicae* 39, no. 3/4 (July–December 1992): 326–30.

Johnson, Phyllis J. and Peter Suedfeld. 'Coping with Stress through the Microcosms of Home and Family Among Arctic Whalers and Explorers'. *The History of the Family* 1, no. 1 (1996): 41–62.

Jonsson, Stefan. 'Where Humanism Finds Its Ends: Lessons from Pia Arke and Katarina Pirak Sikku on the Difficulty of Narrating the Arctic'. *Studies in Travel Writing* 20, no. 3 (September 2016): 226–36.

Jordanova, Ludmilla. 'Image Matters'. *Historical Journal* 51, no. 3 (2008): 777–91.

Kaalund, Nanna Katrine Lüders. 'From Science in the Arctic to Arctic Science: A Transnational Study of Arctic Travel Narratives, 1818–1883'. PhD thesis, York University, Toronto, 2017.

'What Happened to John Franklin? Danish and British Perspectives from Francis McClintock's Arctic Expedition, 1857–59'. *Journal of Victorian Culture* 25, no. 2 (2020): 300–14.

Kautz, Beth Dolan. 'Spas and Salutary Landscapes: The Geography of Health in Mary Shelley's Rambles in Germany and Italy'. In Gilroy, *Romantic Geographies*, 165–81.

Keighren, Innes M., Charles W. J. Withers, and Bill Bell. *Travels into Print: Exploration, Writing, and Publishing with John Murray, 1773–1859*. Chicago and London: University of Chicago Press, 2015.

Kindleberger, Charles P. *Mariners and Markets*. Hertfordshire: Harvester Wheatsheaf, 1992.

Koivunen, Leila. *Visualizing Africa in Nineteenth-Century British Travel Accounts*. Abingdon: Routledge, 2008.

Korte, Barbara. *English Travel Writing from Pilgrimages to Postcolonial Explorations*, trans. Catherine Matthias. Houndmills, Basingstoke: Macmillan, 2000.

Krans, Michael. 'Writing for an Elsewhere: Author(ity) and Authenticity in the Texts of the First Franklin Expedition (1819–22)'. *Studies in Canadian Literature* 24, no. 1 (1999): 70–92.

Lalonde, Suzanne. 'Canadian Sovereignty over the Northwest Passage: Rights and Responsibilities'. Keynote presented at the Nordic Association of Canadian Studies XI, University of Turku, Finland, August 2015.

Lambert, Andrew. *Franklin: Tragic Hero of Polar Navigation*. London: Faber & Faber, 2009. US edition: *The Gates of Hell: Sir John Franklin's Tragic Quest for the North West Passage* (New Haven:Yale University Press, 2009).

Lanone, Catherine. 'Arctic Spectacles in Jane Eyre and Villette'. *Brontë Studies* 34, no. 2 (July 2009): 117–26.

Leask, Nigel. *Curiosity and the Aesthetics of Travel Writing, 1770–1840*. Oxford: Oxford University Press, 2002.

'"Wandering through Elbis": Absorption and Containment in Romantic Exoticism'. In Fulford and Kitson, *Romanticism and Colonialism*, 165–88.

Leek, Michael E. *The Art of Nautical Illustration*. Royston: Quantum Books, 2002.

Leeuw, Sarah de. 'Alice through the Looking Glass: Emotion, Personal Connection, and Reading Colonial Archives along the Grain'. *Journal of Historical Geography* 38, no. 3 (2012): 273–81.

Lefebvre, Henri. *The Production of Space*, trans. Donald Nicholson-Smith. Oxford: Blackwell, 1991.

Levere, Trevor H. 'Science and the Canadian Arctic, 1818–76, from Sir John Ross to Sir George Strong Nares'. *Arctic* 41, no. 2 (June 1988): 127–37.

Science and the Canadian Arctic: A Century of Exploration, 1818–1918. Cambridge: Cambridge University Press, 1993.

Lightman, Bernard. 'Introduction: Victorian Science and Popular Visual Culture'. *Early Popular Visual Culture* 10, no. 1 (February 2012): 1–5.

'The Visual Theology of Victorian Popularizers of Science: From Reverent Eye to Chemical Retina'. *Isis* 91, no. 4 (Dec 2000): 651–80.

Loomis, Chauncey C. 'The Arctic Sublime'. In *Nature and the Victorian Imagination*, ed. U. C. Knoepflmacher and G. B. Tennyson, 95–112. Berkeley and Los Angeles: University of California Press, 1977.

Lopez, Barry. *Arctic Dreams: Imagination and Desire in a Northern Landscape*. 2nd ed. London: Harvill Press, 1999.

MacLaren, Ian S. 'From Exploration to Publication: The Evolution of a 19th-Century Arctic Narrative'. *Arctic* 47, no. 1 (March 1994): 43–53.

'Retaining Captaincy of the Soul: Response to Nature in the First Franklin Expedition'. *Essays on Canadian Writing* 28 (Spring 1984): 57–92.

Review of *Tracing the Connected Narrative: Arctic Exploration in British Print Culture, 1818–1860*, by Janice Cavell. *Arctic* 63, no. 1 (March 2010): 113–15.

'The Aesthetic Map of the North, 1845–1859'. *Arctic* 38, no. 2 (1985): 89–103.

'The Aesthetic Mapping of Nature in the Second Franklin Expedition'. *Journal of Canadian Studies* 20, no. 1 (1985): 39–58.

Maidment, B. E. *Reading Popular Prints, 1790–1870*. 2nd ed. Manchester: Manchester University Press, 2001.

Markham, Albert Hastings. *The Life of Sir Clements R. Markham, K.C.B., F.R.S.* London: John Murray, 1917.

Martin, Constance. 'Perceptions of Arctic Landscapes in the Art of the British Explorers, 1818–1859'. Master's thesis, University of Calgary, 1981.

'"Toward No Earthly Pole": The Search for Franklin in Nineteenth Century Art'. Inter-Nord 19 (1990): 87–100.

Martins, Luciana. 'The Art of Tropical Travel, 1768–1830'. In *Georgian Geographies: Essays on Space, Place and Landscape in the Eighteenth Century*, ed. Miles Ogborn and Charles W. J. Withers, 72–91. Manchester: Manchester University Press, 2004.

Mays, S., G. J. R. Maat, and H. H. Boer. 'Scurvy as a Factor in the Loss of the 1845 Franklin Expedition to the Arctic: A Reconsideration'. *International Journal of Osteoarchaeology* 25 (2015): 334–44.

McCorristine, Shane. '"Involuntarily We Listen": Hearing the Aurora Borealis in Nineteenth-Century Arctic Exploration and Science'. *Canadian Journal of History* 48 (Spring–Summer 2013): 29–61.

McGoogan, Kenneth. *Fatal Passage: The Untold Story of Scotsman John Rae, the Arctic Adventurer who Discovered the Fate of Franklin*. London: Bantam, 2002.

McGuire, James and James Quinn, eds. *Dictionary of Irish Biography*. Cambridge: Cambridge University Press, 2009. http://dib.cambridge.org.

McKitterick, David, ed. *The Cambridge History of the Book in Britain*. Vol. 6 *1830–1914*. Cambridge: Cambridge University Press, 2009.

McNamee, Robert et al., eds. *Electronic Enlightenment Scholarly Edition of Correspondence*. Oxford: University of Oxford, 2008–17. www.e-enlightenment.com.

Merrit, Melissa McBay. 'The Moral Source of the Kantian Sublime'. In Costello, *The Sublime*, 37–49.

Millar, Delia. 'William Simpson (1823–1899)'. In *Oxford Dictionary of National Biography*. Oxford University Press, 2004. www.oxforddnb.com/view/article/25597.

Mills, Sara. 'Written on the Landscape: Mary Wollstonecraft's *Letters Written during a Short Residence in Sweden, Norway and Denmark*'. In Gilroy, *Romantic Geographies:* 19–34.

Mills, William J. *Exploring Polar Frontiers: A Historical Encyclopedia*. Vol. 1, *A – M*. Santa Barbara: ABC-CLIO, 2003.

Mitchell, W. J. T., ed. *Iconology: Image, Text, Ideology*. Chicago: University of Chicago Press, 1986.

Landscape and Power. Chicago: University of Chicago Press, 2002.

Moretti, Franco. *Graphs, Maps, Trees*. London: Verso, 2005.

Moss, Sarah. Review of *White Horizon: The Arctic in the Nineteenth-Century British Imagination*, by Jen Hill. *Nineteenth-Century Literature* 63, no.3 (December 2008): 414–17.

Scott's Last Biscuit: The Literature of Polar Travel. Oxford: Signal, 2006.

Neatby, L. H. 'Robert J. L. McClure (1807–1873)'. *Arctic* 38, no. 1 (March 1985): 70–1.

'Samuel Gurney Cresswell (1827–1867)'. *Arctic* 35, no. 4 (December 1982): 554–5.

The Search for Franklin. Worthing: Littlehampton Book Services, 1970.

Nicolson, Marjorie Hope. *Mountain Gloom, Mountain Glory*. Seattle: University of Washington Press, 1997.

Noakes, Richard. 'Science in Mid-Victorian Punch'. *Endeavour* 26, no. 3 (2002): 92–6.

O'Dochartaigh, Eavan. 'Visual Representation of the Arctic and Public Spectacle in Mid-Victorian England: A Study of the Arctic Art of WHJ Browne'. Master's thesis, University of Cambridge, 2004.

O'Neill, Patrick B. 'Theatre in the North: Staging Practices of the British Navy in the Canadian Arctic'. *Dalhousie Review* 74, no. 3 (1994): 356–84.

Østergård, Else. *Woven into the Earth: Textiles from Norse Greenland*. Aarhus: Aarhus University Press, 2004.

Paden, Roger. 'Picturesque Landscape Painting and Environmental Aesthetics'. *Journal of Aesthetic Education* 49, no. 2 (Summer 2015): 39–61.

Palinkas, Lawrence A. and Peter Suedfeld. 'Psychological Effects of Polar Expeditions'. *Lancet* 371 (January 2008): 153–63.

Panofsky, Erwin. *Studies in Iconology: Humanistic Themes in the Art of the Renaissance*. Boulder: Westview Press, 1972.

Payton, Brian. *The Ice Passage: A True Story of Ambition, Disaster, and Endurance in the Arctic Wilderness*. Toronto: Anchor Canada, 2010.

Peck, Robert M. 'The Art of the Arctic: British Painting in the Far North'. *Journal for Maritime Research* 14, no. 2 (2012): 67–93.

Pelto, Mauri. 'Upernavik Glacier Retreat, NW Greenland'. *From a Glaciers [sic] Perspective* (blog). https://glacierchange.wordpress.com/2011/09/17/upernavik-glacier-retreat-nw-greenland/.

Pesso-Miquel, Catherine. '"In the Company of Strangers": Shedding Light on Robert McClure's Claim of Discovery (1850–7)'. In Regard, *Arctic Exploration*, 61–77.

Pettersson, Lennart. 'Some Aspects on the Pictures of the North'. *Nordlit* 23 (2008): 251–72.

Piola, Erika. 'The Rise of Early American Lithography and Antebellum Visual Culture'. *Winterthur Portfolio* 48, no. 2/3 (2014): 125–38.

Pointon, Marcia. 'Geology and Landscape Painting in Nineteenth-Century England'. In *Images of the Earth*, ed. Ludmilla Jordanova and Roy Porter, 93–123. 2nd ed. London: British Society for the History of Science, 1997.

Pomereu, Jean de. 'The Exploration of Indlandsis: A Cultural and Scientific History of Ice Sheets to 1970'. PhD thesis, University of Exeter, 2015.

Potkay, Adam. 'The British Romantic Sublime'. In Costello, *The Sublime*, 203–16.

Potter, Russell. *Arctic Spectacles: The Frozen North in Visual Culture 1818–1875*. Seattle and London: University of Washington Press, 2007.

Finding Franklin: The Untold Story of a 165-Year Search. Montreal and Kingston: McGill-Queen's University Press, 2016.

'Icebergs at Vauxhall'. *Victorian Review* 36, no. 2 (Fall 2010): 27–31.

'Introduction: Exploration and Sacrifice: The Cultural Logic of Arctic Discovery'. In Regard, *Arctic Exploration*, 1–17.

Potter, Russell and Douglas W. Wamsley. 'The Sublime yet Awful Grandeur: the Arctic Panoramas of Elisha Kent Kane'. *Polar Record* 35 (1999): 193–206.

Poulsom, Neville W. *British Polar Exploration and Research: A Historical and Medallic Record with Biographies 1818–1999*. London: Savannah Publications, 2000.

Powell, Richard C. 'Inscribing the North: The Historical Geographies of Circumpolar Description'. Paper presented at the International Conference of Historical Geographers, Royal Geographical Society, London, July 2015.

Review of *The Last Imaginary Place: A Human History of the Arctic World*, by Robert McGhee. *Canadian Historical Review* 87, no. 2 (June 2006): 332–4.

Pratt, Mary Louise. *Imperial Eyes: Travel Writing and Transculturation*. London: Routledge, 1992.

Preziosi, Donald, ed. *The Art of Art History: A Critical Anthology*. Oxford: Oxford University Press, 1998.

Qikiqtani Inuit Association. *Qikiqtani Truth Commission: Community Histories 1950–1975: Resolute Bay* (Iqaluit: Inhabit Media, 2013).

Quilley, Geoff. '"By Cruel Foes Oppress'd": British Naval Draughtsmen in Tahiti and the South Pacific in the 1840s'. *Journal of Historical Geography* 43 (January 2014): 71–84.

Empire to Nation: Art, History and the Visualization of Maritime Britain, 1768–1829. New Haven and London: Yale University Press, 2011.

'Introduction: Mapping the Art of Travel and Exploration'. *Journal of Historical Geography* 43 (January 2014): 2–8.

'Re-enacting Art and Travel'. *Tate Papers* 17 (May 2012). http://www.tate.org.uk/research/publications/tate-papers/re-enacting-art-and-travel.

Qureshi, Sadiah. *Peoples on Parade: Exhibition, Empire, and Anthropology in Nineteenth-Century Britain*. Chicago: University of Chicago Press, 2011.

Rediker, Marcus. *Between the Devil and the Deep Blue Sea.*, Canto ed. Cambridge: Cambridge University Press, 1993.

Regard, Frédéric, ed. *Arctic Exploration in the Nineteenth Century*. London: Pickering and Chatto, 2013.

Richards, Sys Birgit. 'The Art of Nineteenth-Century Explorer-Artists in the Arctic Regions of North America'. Master's thesis, University of Calgary, 1994.

Richardson, Alan. *The Neural Sublime: Cognitive Theories and Romantic Texts*. Baltimore: Johns Hopkins University Press, 2010.

Ridpath, Ian. *A Dictionary of Astronomy*. Oxford: Oxford University Press, 2012.

Robinson, Michael. 'Manliness and Exploration: The Discovery of the North Pole'. *OSIRIS* 30 (2015): 89–109.

Rockett, Kevin and Emer Rockett. *Magic Lantern, Panorama and Moving Picture Shows in Ireland, 1786–1909*. Dublin: Four Courts Press, 2011.

Rose, Gillian. *Feminism and Geography: The Limits of Geographical Knowledge*. Cambridge: Polity, 1993.

Ross, W. Gillies 'The Type and Number of Expeditions in the Franklin Search 1847–1859'. *Arctic* 55, no. 1 (March 2002): 57–69.

Rozwadowski, Helen M. 'Small World: Forging a Scientific Maritime Culture for Oceanography'. *Isis* 87 (2005): 409–29.

Rudwick, Martin J. S. 'The Emergence of a Visual Language for Geological Science 1760–1840'. *History of Science* 14 (1976): 149–95.

Ryall, Anka. 'The Arctic Playground of Europe: Sir Martin Conway's Svalbard'. *Nordlit* 35 (2015): 29–45.

Sandler, Martin W. *Resolute: The Epic Search for the Northwest Passage and John Franklin, and the Discovery of the Queen's Ghost Ship*. New York: Sterling, 2008.

Said, Edward. *Orientalism*. London: Penguin, 1985.

The World, the Text, and the Critic. Cambridge, MA: Harvard University Press, 1983.

Savours, Ann. Review of *Northwest Passage: The Quest for an Arctic Route to the East*, by Edward Struzik. *Geographical Journal* 158, no. 3 (November 1992): 328.

The Search for the North West Passage. London: Chatham Publishing, 1999.

Sawin, Mark Metzler. 'Raising Kane: Elisha Kent Kane and the Culture of Fame in Antebellum America'. *Transactions of the American Philosophical Society* 98, no. 3 (2008): 1–368.

Schama, Simon. *Landscape and Memory*. London: HarperCollins, 1995.

Schultheiss, Katrin. 'The Ends of the Earth and the "Heroic Age" of Polar Exploration: A Review Essay'. *Historically Speaking* 10, no. 2 (April 2009): 14–17.

Sedgwick, Eve Kosofsky. *Epistemology of the Closet*. 2nd ed. Berkeley and Los Angeles: University of California Press, 2008.

Sera-Shriar, Efram. 'Anthropometric Portraiture and Victorian Anthropology: Situating Francis Galton's Photographic Work in the Late 1870s'. *History of Science* 53, no. 2 (2015): 155–79.

'Arctic Observers: Richard King, Monogenism and the Historicisation of Inuit through Travel Narratives'. *Studies in History and Philosophy of Biological and Biomedical Sciences* 51 (June 2015): 23–31.

The Making of British Anthropology, 1813–1871. London: Pickering and Chatto, 2013.

Sha, Richard C. *The Visual and Verbal Sketch in British Romanticism*. Philadelphia: University of Pennsylvania Press, 1998.

Shapin, Steven and Simon Schaffer. *Leviathan and the Air-Pump: Hobbes, Boyle, and the Experimental Life*. Princeton: Princeton University Press, 1989.

Sharp, Joanne P. *Geographies of Postcolonialism: Spaces of Power and Representation*. London: Sage, 2009.

Shelly, James. '18th Century British Aesthetics'. In *Stanford Encyclopedia of Philosophy*, ed. Edward N. Zalta. Stanford University, 1997–. Article revised 18 July 2014. http://plato.stanford.edu/archives/fall2014/entries/aesthetics-18th-british.

Shields, Rob. *Lefebvre, Love and Struggle: Spatial Dialectics*. London: Routledge, 1999.

Places on the Margin: Alternative Geographies of Modernity. London: Routledge, 1991.

Smith, Bernard. *European Vision and the South Pacific*. New Haven: Yale University Press, 1985.

Imagining the Pacific: In the Wake of the Cook Voyages. New Haven and London: Yale University Press, 1992.

Spufford, Francis. *I May Be Some Time: Ice and the English Imagination*. London: Faber & Faber, 1996.

Stacey, Robert. 'From "Icy Picture" to "Extensive Prospect": The Panorama of Rupert's Land and the Far North in the Artist's Eye, 1770–1830'. In *Rupert's Land: A Cultural Tapestry*, ed. Richard C. Davis, 147–89. Waterloo, ON: Wilfrid Laurier University Press, 1988.

Stainton, Lindsay. *Nature into Art: English Landscape Watercolours*. London: British Museum Press, 1991.

Stam, David H. and Deirdre C. Stam. 'Bending Time: The Function of Periodicals in Nineteenth-Century Polar Naval Expeditions'. *Victorian Periodicals Review* 41, no. 4 (Winter 2008): 301–22.

Steinberg, Philip E. 'You Are (Not) Here: On the Ambiguity of Flag Planting and Finger Pointing in the Arctic'. *Political Geography* 29 (2010): 81–4.

Stewart, E. J. and J. Dawson. 'A Matter of Good Fortune? The Grounding of the *Clipper Adventurer* in the Northwest Passage, Arctic Canada'. *Arctic* 64, no. 2 (June 2011): 263–7.

Stillinger, Jack and Deidre Shauna Lynch, eds. *The Romantic Period*. Vol. D, *The Norton Anthology of English Literature*. 8th ed. W.W. Norton: New York and London, 2006.

Strange, Carolyn. 'Reconsidering the "Tragic" Scott Expedition: Cheerful Masculine Home-making in Antarctica, 1910–1913'. *Journal of Social History* 48, no. 1 (Fall 2012): 66–88.

Sullivan, Alvin, ed. *British Literary Magazines: The Victorian and Edwardian Age, 1837–1913*. London: Greenwood Press, 1984.

Sutton, Max Keith. '"Inverse Sublimity" in Victorian Humor'. *Victorian Studies* 10, no. 2 (December 1966): 177–92.

Taichman, R. S., T. Gross, and M. P. Maceachern. 'A Critical Assessment of the Oral Condition of the Crew of the Franklin Expedition'. *Arctic* 70, no. 1 (March 2017): 25–36.

Tekiner, Deniz. *Modern Art and the Romantic Vision*. Lanham: University Press of America, 2000.

Till, Karen. 'Neotraditional Towns and Urban Villages: The Cultural Production of a Geography or "Otherness"'. *Environment and Planning D: Society and Space* 11, no. 6 (1993): 709–32.

Tilley, Elizabeth. 'Periodicals'. In *The Irish Book in English, 1800–1891*, ed. James H. Murphy, 144–70. Vol. 4 of *The Oxford History of the Irish Book*. Oxford: Oxford University Press, 2011.

Thomas, Julia. *Pictorial Victorians: The Inscription of Values in Word and Image*. Athens: University of Ohio Press, 2004.

Thompson, Carl. *The Suffering Traveller and the Romantic Imagination*. Oxford: Clarendon Press, 2007.

Travel Writing. London: Routledge, 2011.

Thompson,Carl, ed. *The Routledge Companion to Travel Writing*. London: Routledge, 2016.

Topping, Margaret. 'Travel Writing and Visual Culture', in Thompson, *Routledge Companion to Travel Writing*, 78–88.

Törnvall, Gunilla. 'From Copperplate to Color Lithograph: On the Modernisation of an Illustrated Flora 1800–1900'. *Book History* 20 (2017): 126–49.

Townsend, Dabney. 'The Picturesque'. *The Journal of Aesthetics and Art Criticism* 55, no. 4 (October 1997): 365–76.

Trettin, H. P., ed. *Geology of the Innuitian Orogen and Arctic Platform of Canada and Greenland*. Ottawa: Geological Survey of Canada, 1991.

Tuan, Yi-Fu. *Space and Place: The Perspective of Experience*. Minneapolis: University of Minnesota Press, 1977.

Topophilia: A Study of Environmental Perception, Attitudes, and Values. Morningside ed. New York: Columbia University Press, 1990.

Twyman, Michael. *Lithography 1800-1850: The Techniques of Drawing on Stone in England and France and Their Application in Works of Topography*. Oxford: Oxford University Press, 1970.

'The Illustration Revolution'. In McKitterick, *The Cambridge History of the Book in Britain*, vol. 6, 117–43.

Urry, John. *The Tourist Gaze*. London: Sage, 2002.

Vaughan, Richard. *The Arctic: A History*.Stroud: Sutton Publishing, 1999.

Vaughan, William. *Romantic Art*. London: Thames and Hudson, 1988.

Venning, Barry. 'Turner's Whaling Subjects'. *The Burlington Magazine* 127, no. 983 (February 1985): 75–83.

Ventsel, Aimar. '"Our Beer is Better Than Theirs": Sub-Cultural Comics as a Joking Relationship with Identity'. *Folklore: Electronic Journal of Folklore* 46 (2010): 143–60.

Walpole, Garth. *Relics of the Franklin Expedition: Discovering Artifacts from the Doomed Arctic Voyage of 1845*, ed. Russell Potter (Jefferson: McFarland, 2017).

Warner, Marina. 'Imaginary Passages'. *Third Text* 21, no. 3 (September 2007): 487–98.

Once Upon a Time: A Short History of Fairy Tale. Oxford: Oxford University Press, 2014.

Warrior, Claire. '"On Thin Ice": The Polar Displays at the National Maritime Museum, Greenwich'. *Museum History Journal*, 6, no. 1 (November 2013): 56–71.

Watt-Cloutier, Sheila. *The Right to Be Cold: One Woman's Fight to Protect the Arctic and Save the Planet from Climate Change*. Minneapolis: University of Minnesota Press, 2018.

Weiskel, Thomas. *The Romantic Sublime: Studies in the Structure and Psychology of Transcendence*. Baltimore: Johns Hopkins University Press, 1976.

West, Shearer, ed. *The Victorians and Race*. Aldershot: Scolar, 1996.

Wiebe, Rudy. *A Discovery of Strangers*. Toronto: Penguin Random House, 1994.

Wilson, Eric G. *The Spiritual History of Ice: Romanticism, Science, and the Imagination*. Basingstoke: Palgrave Macmillan, 2003.

Wood, Christopher. *Dictionary of British Art*. Vol. 4, bk. 1, of *Victorian Painters*. Woodbridge: Antique Collectors' Club, 1998.

Wylie, John. *Landscape*. London: Routledge, 2007.

Yar, Majid. 'Panoptic Power and the Pathologisation of Vision: Critical Reflections on the Foucauldian Thesis'. *Surveillance and Society* 1, no. 3 (2003): 254–71.

Youngs, Tim. *The Cambridge Introduction to Travel Writing*. Cambridge: Cambridge University Press, 2013.

Youngs, Tim, ed. *Travel Writing in the Nineteenth Century: Filling in the Blank Spaces*. London: Anthem Press, 2006.

Index

CAMBRIDGE STUDIES IN NINETEENTH-CENTURY LITERATURE AND CULTURE

GENERAL EDITORS
Kate Flint, *University of Southern California*
Clare Pettitt, *King's College London*

Titles published

1. *The Sickroom in Victorian Fiction: The Art of Being Ill*
   MIRIAM BAILIN, *Washington University*
2. *Muscular Christianity: Embodying the Victorian Age*
   edited by DONALD E. HALL, *California State University, Northridge*
3. *Victorian Masculinities: Manhood and Masculine Poetics in Early Victorian Literature and Art*
   HERBERT SUSSMAN, *Northeastern University, Boston*
4. *Byron and the Victorians*
   ANDREW ELFENBEIN, *University of Minnesota*
5. *Literature in the Marketplace: Nineteenth-Century British Publishing and the Circulation of Books*
   edited by JOHN O. JORDAN, *University of California,* SANTA CRUZ and ROBERT L. PATTEN, *Rice University, Houston*
6. *Victorian Photography, Painting and Poetry*
   LINDSAY SMITH, *University of Sussex*
7. *Charlotte Brontë and Victorian Psychology*
   SALLY SHUTTLEWORTH, *University of Sheffield*
8. *The Gothic Body: Sexuality, Materialism and Degeneration at the Fin de Siècle*
   KELLY HURLEY, *University of Colorado at Boulder*
9. *Rereading Walter Pater*
   WILLIAM F. SHUTER, *Eastern Michigan University*
10. *Remaking Queen Victoria*
    edited by MARGARET HOMANS, *Yale University* and ADRIENNE MUNICH, *State University of New York, Stony Brook*
11. *Disease, Desire, and the Body in Victorian Women's Popular Novels*
    PAMELA K. GILBERT, *University of Florida*
12. *Realism, Representation, and the Arts in Nineteenth-Century Literature*
    ALISON BYERLY, *Middlebury College, Vermont*
13. *Literary Culture and the Pacific*
    VANESSA SMITH, *University of Sydney*
14. *Professional Domesticity in the Victorian Novel: Women, Work and Home*
    MONICA F. COHEN
15. *Victorian Renovations of the Novel: Narrative Annexes and the Boundaries of Representation*
    SUZANNE KEEN, *Washington and Lee University, Virginia*

16. *Actresses on the Victorian Stage: Feminine Performance and the Galatea Myth*
GAIL MARSHALL, *University of Leeds*
17. *Death and the Mother from Dickens to Freud: Victorian Fiction and the Anxiety of Origin*
CAROLYN DEVER, *Vanderbilt University, Tennessee*
18. *Ancestry and Narrative in Nineteenth-Century British Literature: Blood Relations from Edgeworth to Hardy*
SOPHIE GILMARTIN, *Royal Holloway, University of London*
19. *Dickens, Novel Reading, and the Victorian Popular Theatre*
DEBORAH VLOCK
20. *After Dickens: Reading, Adaptation and Performance*
JOHN GLAVIN, *Georgetown University, Washington DC*
21. *Victorian Women Writers and the Woman Question*
edited by NICOLA DIANE THOMPSON, *Kingston University, London*
22. *Rhythm and Will in Victorian Poetry*
MATTHEW CAMPBELL, *University of Sheffield*
23. *Gender, Race, and the Writing of Empire: Public Discourse and the Boer War*
PAULA M. KREBS, *Wheaton College, Massachusetts*
24. *Ruskin's God*
MICHAEL WHEELER, *University of Southampton*
25. *Dickens and the Daughter of the House*
HILARY M. SCHOR, *University of Southern California*
26. *Detective Fiction and the Rise of Forensic Science*
RONALD R. THOMAS, *Trinity College, Hartford, Connecticut*
27. *Testimony and Advocacy in Victorian Law, Literature, and Theology*
JAN-MELISSA SCHRAMM, *Trinity Hall, Cambridge*
28. *Victorian Writing about Risk: Imagining a Safe England in a Dangerous World*
ELAINE FREEDGOOD, *University of Pennsylvania*
29. *Physiognomy and the Meaning of Expression in Nineteenth-Century Culture*
LUCY HARTLEY, *University of Southampton*
30. *The Victorian Parlour: A Cultural Study*
THAD LOGAN, *Rice University, Houston*
31. *Aestheticism and Sexual Parody 1840–1940*
DENNIS DENISOFF, *Ryerson University, Toronto*
32. *Literature, Technology and Magical Thinking, 1880–1920*
PAMELA THURSCHWELL, *University College London*
33. *Fairies in Nineteenth-Century Art and Literature*
NICOLA BOWN, *Birkbeck College, University of London*
34. *George Eliot and the British Empire*
NANCY HENRY, *The State University of New York, Binghamton*
35. *Women's Poetry and Religion in Victorian England: Jewish Identity and Christian Culture*
CYNTHIA SCHEINBERG, *Mills College, California*
36. *Victorian Literature and the Anorexic Body*
ANNA KRUGOVOY SILVER, *Mercer University, Georgia*

37. *Eavesdropping in the Novel from Austen to Proust*
ANN GAYLIN, *Yale University*
38. *Missionary Writing and Empire, 1800–1860*
ANNA JOHNSTON, *University of Tasmania*
39. *London and the Culture of Homosexuality, 1885–1914*
MATT COOK, *Keele University*
40. *Fiction, Famine, and the Rise of Economics in Victorian Britain and Ireland*
GORDON BIGELOW, *Rhodes College, Tennessee*
41. *Gender and the Victorian Periodical*
HILARY FRASER, *Birkbeck College, University of London*, JUDITH JOHNSTON and STEPHANIE GREEN, *University of Western Australia*
42. *The Victorian Supernatural*
edited by NICOLA BOWN, *Birkbeck College, University of London*, CAROLYN BURDETT, *London Metropolitan University*, and PAMELA THURSCHWELL, *University College London*
43. *The Indian Mutiny and the British Imagination*
GAUTAM CHAKRAVARTY, *University of Delhi*
44. *The Revolution in Popular Literature: Print, Politics and the People*
IAN HAYWOOD, *Roehampton University of Surrey*
45. *Science in the Nineteenth-Century Periodical: Reading the Magazine of Nature*
GEOFFREY CANTOR, *University of Leeds*, GOWAN DAWSON, *University of Leicester*, GRAEME GOODAY, *University of Leeds*, RICHARD NOAKES, *University of Cambridge*, SALLY SHUTTLEWORTH, *University of Sheffield*, and JONATHAN R. TOPHAM, *University of Leeds*
46. *Literature and Medicine in Nineteenth-Century Britain from Mary Shelley to George Eliot*
JANIS MCLARREN CALDWELL, *Wake Forest University*
47. *The Child Writer from Austen to Woolf*
edited by CHRISTINE ALEXANDER, *University of New South Wales*, and JULIET MCMASTER, *University of Alberta*
48. *From Dickens to Dracula: Gothic, Economics, and Victorian Fiction*
GAIL TURLEY HOUSTON, *University of New Mexico*
49. *Voice and the Victorian Storyteller*
IVAN KREILKAMP, *University of Indiana*
50. *Charles Darwin and Victorian Visual Culture*
JONATHAN SMITH, *University of Michigan-Dearborn*
51. *Catholicism, Sexual Deviance, and Victorian Gothic Culture*
PATRICK R. O'MALLEY, *Georgetown University*
52. *Epic and Empire in Nineteenth-Century Britain*
SIMON DENTITH, *University of Gloucestershire*
53. *Victorian Honeymoons: Journeys to the Conjugal*
HELENA MICHIE, *Rice University*
54. *The Jewess in Nineteenth-Century British Literary Culture*
NADIA VALMAN, *University of Southampton*

55. *Ireland, India and Nationalism in Nineteenth-Century Literature*
JULIA WRIGHT, *Dalhousie University*
56. *Dickens and the Popular Radical Imagination*
SALLY LEDGER, *Birkbeck College, University of London*
57. *Darwin, Literature and Victorian Respectability*
GOWAN DAWSON, *University of Leicester*
58. *'Michael Field': Poetry, Aestheticism and the Fin de Siècle*
MARION THAIN, *University of Birmingham*
59. *Colonies, Cults and Evolution: Literature, Science and Culture in Nineteenth-Century Writing*
DAVID AMIGONI, *Keele University*
60. *Realism, Photography and Nineteenth-Century Fiction*
DANIEL A. NOVAK, *Louisiana State University*
61. *Caribbean Culture and British Fiction in the Atlantic World, 1780–1870*
TIM WATSON, *University of Miami*
62. *The Poetry of Chartism: Aesthetics, Politics, History*
MICHAEL SANDERS, *University of Manchester*
63. *Literature and Dance in Nineteenth-Century Britain: Jane Austen to the New Woman*
CHERYL WILSON, *Indiana University*
64. *Shakespeare and Victorian Women*
GAIL MARSHALL, *Oxford Brookes University*
65. *The Tragi-Comedy of Victorian Fatherhood*
VALERIE SANDERS, *University of Hull*
66. *Darwin and the Memory of the Human: Evolution, Savages, and South America*
CANNON SCHMITT, *University of Toronto*
67. *From Sketch to Novel: The Development of Victorian Fiction*
AMANPAL GARCHA, *Ohio State University*
68. *The Crimean War and the British Imagination*
STEFANIE MARKOVITS, *Yale University*
69. *Shock, Memory and the Unconscious in Victorian Fiction*
JILL L. MATUS, *University of Toronto*
70. *Sensation and Modernity in the 1860s*
NICHOLAS DALY, *University College Dublin*
71. *Ghost-Seers, Detectives, and Spiritualists: Theories of Vision in Victorian Literature and Science*
SRDJAN SMAJIĆ, *Furman University*
72. *Satire in an Age of Realism*
AARON MATZ, *Scripps College, California*
73. *Thinking About Other People in Nineteenth-Century British Writing*
ADELA PINCH, *University of Michigan*
74. *Tuberculosis and the Victorian Literary Imagination*
KATHERINE BYRNE, *University of Ulster, Coleraine*

75. *Urban Realism and the Cosmopolitan Imagination in the Nineteenth Century: Visible City, Invisible World*
TANYA AGATHOCLEOUS, *Hunter College, City University of New York*
76. *Women, Literature, and the Domesticated Landscape: England's Disciples of Flora, 1780–1870*
JUDITH W. PAGE, *University of Florida*, and ELISE L. SMITH, *Millsaps College, Mississippi*
77. *Time and the Moment in Victorian Literature and Society*
SUE ZEMKA, *University of Colorado*
78. *Popular Fiction and Brain Science in the Late Nineteenth Century*
ANNE STILES, *Washington State University*
79. *Picturing Reform in Victorian Britain*
JANICE CARLISLE, *Yale University*
80. *Atonement and Self-Sacrifice in Nineteenth-Century Narrative*
JAN-MELISSA SCHRAMM, *University of Cambridge*
81. *The Silver Fork Novel: Fashionable Fiction in the Age of Reform*
EDWARD COPELAND, *Pomona College, California*
82. *Oscar Wilde and Ancient Greece*
IAIN ROSS, *Colchester Royal Grammar School*
83. *The Poetry of Victorian Scientists: Style, Science and Nonsense*
DANIEL BROWN, *University of Southampton*
84. *Moral Authority, Men of Science, and the Victorian Novel*
ANNE DEWITT, *Princeton Writing Program*
85. *China and the Victorian Imagination: Empires Entwined*
ROSS G. FORMAN, *University of Warwick*
86. *Dickens's Style*
edited by DANIEL TYLER, *University of Oxford*
87. *The Formation of the Victorian Literary Profession*
RICHARD SALMON, *University of Leeds*
88. *Before George Eliot: Marian Evans and the Periodical Press*
FIONNUALA DILLANE, *University College Dublin*
89. *The Victorian Novel and the Space of Art: Fictional Form on Display*
DEHN GILMORE, *California Institute of Technology*
90. *George Eliot and Money: Economics, Ethics and Literature*
DERMOT COLEMAN, *Independent Scholar*
91. *Masculinity and the New Imperialism: Rewriting Manhood in British Popular Literature, 1870–1914*
BRADLEY DEANE, *University of Minnesota*
92. *Evolution and Victorian Culture*
edited by BERNARD LIGHTMAN, *York University, Toronto*, and BENNETT ZON, *University of Durham*
93. *Victorian Literature, Energy, and the Ecological Imagination*
ALLEN MACDUFFIE, *University of Texas, Austin*
94. *Popular Literature, Authorship and the Occult in Late Victorian Britain*
ANDREW MCCANN, *Dartmouth College, New Hampshire*

95. *Women Writing Art History in the Nineteenth Century: Looking Like a Woman*
HILARY FRASER, *Birkbeck College, University of London*
96. *Relics of Death in Victorian Literature and Culture*
DEBORAH LUTZ, *Long Island University, C. W. Post Campus*
97. *The Demographic Imagination and the Nineteenth-Century City: Paris, London, New York*
NICHOLAS DALY, *University College Dublin*
98. *Dickens and the Business of Death*
CLAIRE WOOD, *University of York*
99. *Translation as Transformation in Victorian Poetry*
ANNMARIE DRURY, *Queens College, City University of New York*
100. *The Bigamy Plot: Sensation and Convention in the Victorian Novel*
MAIA MCALEAVEY, *Boston College, Massachusetts*
101. *English Fiction and the Evolution of Language, 1850–1914*
WILL ABBERLEY, *University of Oxford*
102. *The Racial Hand in the Victorian Imagination*
AVIVA BRIEFEL, *Bowdoin College, Maine*
103. *Evolution and Imagination in Victorian Children's Literature*
JESSICA STRALEY, *University of Utah*
104. *Writing Arctic Disaster: Authorship and Exploration*
ADRIANA CRACIUN, *University of California, Riverside*
105. *Science, Fiction, and the Fin-de-Siècle Periodical Press*
WILL TATTERSDILL, *University of Birmingham*
106. *Democratising Beauty in Nineteenth-Century Britain: Art and the Politics of Public Life*
LUCY HARTLEY, *University of Michigan*
107. *Everyday Words and the Character of Prose in Nineteenth-Century Britain*
JONATHAN FARINA, *Seton Hall University, New Jersey*
108. *Gerard Manley Hopkins and the Poetry of Religious Experience*
MARTIN DUBOIS, *Newcastle University*
109. *Blindness and Writing: From Wordsworth to Gissing*
HEATHER TILLEY, *Birkbeck College, University of London*
110. *An Underground History of Early Victorian Fiction: Chartism, Radical Print Culture, and the Social Problem Novel*
GREGORY VARGO, *New York University*
111. *Automatism and Creative Acts in the Age of New Psychology*
LINDA M. AUSTIN, *Oklahoma State University*
112. *Idleness and Aesthetic Consciousness, 1815–1900*
RICHARD ADELMAN, *University of Sussex*
113. *Poetry, Media, and the Material Body: Autopoetics in Nineteenth-Century Britain*
ASHLEY MILLER, *Albion College, Michigan*
114. *Malaria and Victorian Fictions of Empire*
JESSICA HOWELL, *Texas A&M University*

115. *The Brontës and the Idea of the Human: Science, Ethics, and the Victorian Imagination*
edited by ALEXANDRA LEWIS, *University of Aberdeen*
116. *The Political Lives of Victorian Animals: Liberal Creatures in Literature and Culture*
ANNA FEUERSTEIN, *University of Hawai'i-Manoa*
117. *The Divine in the Commonplace: Recent Natural Histories and the Novel in Britain*
AMY KING, *St John's University, New York*
118. *Plagiarizing the Victorian Novel: Imitation, Parody, Aftertext*
ADAM ABRAHAM, *Virginia Commonwealth University*
119. *Literature, Print Culture, and Media Technologies, 1880–1900: Many Inventions*
RICHARD MENKE, *University of Georgia*
120. *Aging, Duration, and the English Novel: Growing Old from Dickens to Woolf*
JACOB JEWUSIAK, *Newcastle University*
121. *Autobiography, Sensation, and the Commodification of Identity in Victorian Narrative: Life upon the Exchange*
SEAN GRASS, *Rochester Institute of Technology*
122. *Settler Colonialism in Victorian Literature: Economics and Political Identity in the Networks of Empire*
PHILLIP STEER, *Massey University, Auckland*
123. *Mimicry and Display in Victorian Literary Culture: Nature, Science and the Nineteenth-Century Imagination*
WILL ABBERLEY, *University of Sussex*
124. *Victorian Women and Wayward Reading: Crises of Identification*
MARISA PALACIOS KNOX, *University of Texas Rio Grande Valley*
125. *The Victorian Cult of Shakespeare: Bardology in the Nineteenth Century*
CHARLES LAPORTE, *University of Washington*
126. *Children's Literature and the Rise of 'Mind Cure': Positive Thinking and Pseudo-Science at the Fin de Siècle*
ANNE STILES, *Saint Louis University, Missouri*
127. *Virtual Play and the Victorian Novel: The Ethics and Aesthetics of Fictional Experience*
TIMOTHY GAO, *Nanyang Technological University*
128. *Colonial Law in India and the Victorian Imagination*
LEILA NETI, *Occidental College, Los Angeles*
129. *Convalescence in the Nineteenth-Century Novel: The Afterlife of Victorian Illness*
HOSANNA KRIENKE, *University of Wyoming*
130. *Stylistic Virtue and Victorian Fiction: Form, Ethics and the Novel*
MATTHEW SUSSMAN, *The University of Sydney*
131. *Scottish Women's Writing in the Long Nineteenth Century: The Romance of Everyday Life*
JULIET SHIELDS, *University of Washington*

132. *Reimagining Dinosaurs in Late Victorian and Edwardian Literature: How the 'Terrible Lizard' Became a Transatlantic Cultural Icon*
RICHARD FALLON, *The University of Birmingham*

133. *Decadent Ecology in British Literature and Art, 1860–1910: Decay, Desire, and the Pagan Revival*
DENNIS DENISOFF, *University of Tulsa*

134. *Vagrancy in the Victorian Age: Representing the Wandering Poor in Nineteenth-Century Literature and Culture*
ALISTAIR ROBINSON, *New College of the Humanities*

135. *Collaborative Writing in the Long Nineteenth Century: Sympathetic Partnerships and Artistic Creation*
HEATHER BOZANT WITCHER, *Auburn University, Montgomery*

136. *Visual Culture and Arctic Voyages: Personal and Public Art and Literature of the Franklin Search Expeditions*
EAVAN O'DOCHARTAIGH, *National University of Ireland Galway*